LeedVisual GA V3

"A picture is worth a thousand words"

A visual explanation of Green Building basics &
a study guide for LEED Green Associate exam.

www.LeedVisual.com
E-mail - Mail@LeedVisual.com

Printed in the United States of America
First edition
ISBN 978-0-615-33280-2

This book is printed on FSC certified paper.

1 Green Building

2 Leadership in Energy & Environmental Design

3 Sustainable Site

4 Water Efficiency

5 Energy & Atmosphere

6 Material & Resources

7 Indoor Environmental Quality

8 Innovation in Design

9 Regional Priority

How this book is organized

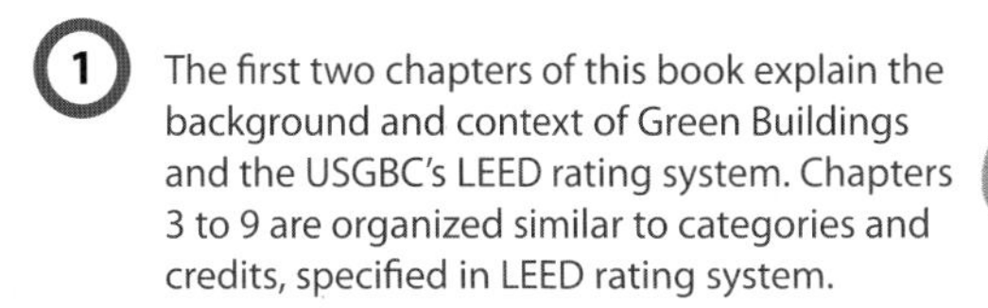

1 The first two chapters of this book explain the background and context of Green Buildings and the USGBC's LEED rating system. Chapters 3 to 9 are organized similar to categories and credits, specified in LEED rating system.

2 Chapters are color coded for easy navigation through the book.

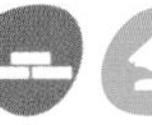

3 

Page headings are named similar to LEED Credits titles.

4 Each credit is divided into 'intent' and 'approach'. The **intent** addresses the environmental issue, and the **approach** explains strategies to solve it.

5 Illustrations are located typically below explanatory text, accompanied by captions. The illustrations are intended to help the reader visualize, what is described in the text. Prior reading of the text, can help in better understanding of sketches and easy memorization of concepts.

6 Glossary relevant to the topic is placed at bottom of the page.

The building design, construction, and operations industry have a profound impact on our natural environment, economy, health, and productivity.

In the United States alone, buildings account for:

- 72% of electricity consumption;
- 39% of energy use;
- 38% of all carbon dioxide (CO_2) emissions;
- 40% of raw materials use;
- 30% of waste output (136 million tons annually); and
- 14% of potable water consumption. [1]

This enormous influence of the built environment, makes it important for us to take action to reduce its impact.

Green Building practices can reduce or even eliminate negative environmental impacts through high performance, market leading design, construction and operation practices. Additionally, this can reduce operation costs, enhance building marketability, increase productivity, and reduce liabilities due to indoor air quality problems.

GREEN BUILDING BASICS

What is Green Building?

Green building is the practice of creating structures and using processes that are environmentally responsible and resource-efficient throughout a building's life-cycle from designing, construction, operation, maintenance, renovation and deconstruction. This practice expands and complements the classical building design concerns of economy, utility, durability, and comfort. Green building is also known as a sustainable or high performance building.

Green buildings are designed to reduce the overall impact of the built environment on human health and the natural environment by:

- Efficiently using energy, water, and other resources
- Protecting occupant health and improving employee productivity
- Reducing waste, pollution and environmental degradation. [2]

Benefits of Green Building

1) Environmental benefits:

- Enhances and protects ecosystems and biodiversity
- Improves air and water quality
- Reduces solid waste
- Conserves natural resources

2) Economic benefits:

- Reduces operating costs
- Enhances asset value and profits
- Improves employee productivity and satisfaction
- Optimizes life-cycle economic performance

3) Health and community benefits:

- Improves air, thermal, and acoustic environments
- Enhances occupant comfort and health
- Minimizes strain on local infrastructure
- Contributes to overall quality of life. [3]

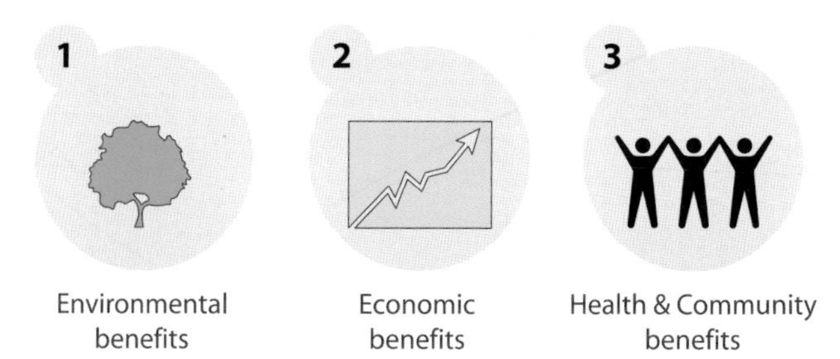

Environmental benefits | Economic benefits | Health & Community benefits

○ ***Carbon footprint:***

Carbon footprint is a measure of greenhouse gas emissions caused by an activity.

What is meant by Triple Bottom Line?

Triple Bottom Line (TBL or 3BL, also know as people, planet, profit) was a phrase coined by John Elkington. It is a criterion for measuring organizational success : economical, ecological and social.

USGBC has adapted the triple bottom line to establish metrics and rating systems to measure and recognize building projects based on their performance in the three corresponding dimensions of sustainabilty; society, environment, and economy.

Projects certified under the LEED rating systems demonstrate, that they have addressed elements that balance and enhance all three areas of the triple bottom line, all the three dimensions of sustainability. [4]

ECONOMIC PROSPERITY
ENVIRONMENTAL STEWARDSHIP
SOCIAL RESPONSIBILITY

Economic Prosperity

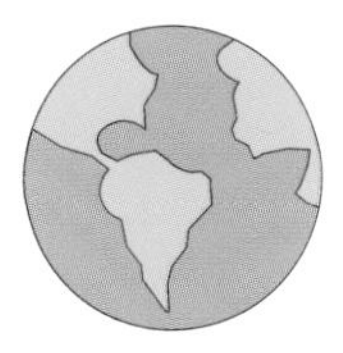

Environmental stewardship

Social Responsibility

What are the types of costs?

There are two types of costs involved in any building project:

1) **Hard Cost** - Price of land, building cost, cost of construction, etc.
2) **Soft Cost** - Consultants fee, drawings fee, interest payments, city permits, legal fee, etc.

Green buildings also consider life cycle costs. NIST (The National Institute of Standards and Technology) defines Life Cycle Cost (LCC) as the total discounted dollar cost of owning, operating, maintaining, and disposing of a building or a building system over a period of time.

Life Cycle Cost Analysis (LCCA) is an economic evaluation technique that determines the total cost of owning and operating a facility over a period of time. [5]

Studies have shown that there is no significant difference in the average cost of green buildings as compared to non-green buildings. However, following are the primary reasons that a green or LEED certified building would be more expensive:

1) Green building goals were considered as an add-on after the building was completed.
2) Lack of an integrated design team or abscence of communication between team members.
3) Value engineering some of the green building goals , which cuts down initial cost, but results in higher operation cost.

What is whole building design?

The goal of 'Whole Building' Design is to create a successful high-performance building by applying an integrated design and team approach to a project during the planning and programming phases.

Synergies among credits and sustainability goals yield a successful Whole Building Design. Whole Building Design consists of two components:
1) Integrated design approach
2) Integrated team process. [6]

1: Integrated Design Approach

The 'integrated' design approach asks all members of the building stakeholder community, and the technical planning, design, and construction team to look at the project from many different perspectives. This approach is a deviation from the typical planning and design process of relying on the expertise of specialists who work in their respective specialties somewhat isolated from each other.

Design Objectives of Whole Building Design
In buildings, to achieve a truly successful holistic project, following design objectives must be considered in concert with each other:
- Accessibility
- Aesthetics
- Cost-Effectiveness
- Functionality/Operation
- Historic Preservation
- Productivity
- Security/Safety
- Sustainability. [7]

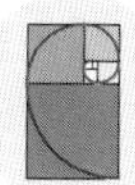

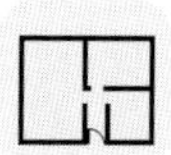

2: Integrated Team Process

Whole Building design in practice also requires an integrated team process in which the design team and all affected stakeholders work together throughout the project phases to evaluate the design for: cost, quality-of-life, future flexibility, efficiency, overall environmental impact, productivity, creativity, and how the occupants will be enlivened.

To create a successful green building, an interactive approach to the design process is required. It means all the stakeholders: everyone involved in the planning, design, use, construction, operation, and maintenance must fully understand the issues and concerns of all the other parties and interact closely throughout the project. [7]

○ ***Design Charrette:***
A design charrette is a focused and collaborative brainstorming session held at the beginning of a project. It encourages an exchange of ideas and information and allows truly integrated design solutions to take form.

After the formation of the U.S. Green Building Council in 1993, the organization's members quickly realized that the sustainable building industry needed a system to define and measure "green buildings". USGBC began to research existing green building metrics and rating systems. Less than a year after formation, the members acted on the initial findings by establishing a committee to focus solely on this topic. The composition of the committee was diverse; it included architects, real estate agents, a building owner, a lawyer, an environmentalist, and industry representatives. This cross section of people and professions added a richness and depth to both the process and to the ultimate product. [1]

Here is a brief time line of the LEED Rating system:
LEED Version 1.0 - The first LEED Pilot Project Program, was launched in August 1998.
LEED Version 2.0 – released in 2000
LEED Version 2.1 – released in 2002
LEED Version 2.2 – released in 2005
LEED Version 3.0 – released in 2009

This chapter explains LEED - its rating systems and certification process.

About USGBC

Established in 1993, Washington, DC based, U.S. Green Building Council (USGBC) is a 501 (c)(3) non-profit organization committed to a prosperous and sustainable future for our nation through cost-efficient and energy-saving green buildings. The members of USGBC represent more than 15,000 organizations across the industry. Their members include building owners and end users, real estate developers, facility managers, architects, designers, engineers, general contractors, subcontractors, product and building system manufacturers and government agencies. [2]

USGBC's Mission:

"To transform the way buildings and communities are designed, built and operated, enabling an environmentally and socially responsible, healthy, and prosperous environment that improves the quality of life." [2]

USGBC's Vision:

"Buildings and communities will regenerate and sustain the health and vitality of all life within a generation." [2]

Chapter Programs

There are 78 regional USGBC chapters nationwide providing green building resources, education, and networking opportunities in their communities. [2]

Education

USGBC provides top quality educational programs on green design, construction, and operations for professionals from all sectors of the building industry. Its online portal to green building education is *www.greenbuild365.org*. USGBC also hosts Green build, the largest international conference and expo focused on green building. [2]

Emerging Green Builders

USGBC's Emerging Green Builders program provides educational opportunities and resources to students and young professionals with the goal of integrating these future leaders into the green building movement. [2]

How to use USGBC in text?

The official organization name is the U.S. Green Building Council. "USGBC" is the official acronym. On any first reference; use the complete name. Subsequent references may use just USGBC.

Following are acceptable examples;

U.S. Green Building Council
USGBC

Unacceptable examples are;

U.S.G.B.C.
U.S. GBC
United States Green Building Council
US Green Building Council
GBC

The USGBC Logo can be used at following places:

- Next to the text discussing LEED Rating System
- Only to refer to the USGBC program or USGBC products
- May not be used to indicate any kind of endorsement by USGBC of any product or service. [2]

How to use LEED in text?

LEED, as a registered trademark of the U.S. Green Building Council, does not have to be spelled out in the first reference; however, when applying the use of the complete name, "Leadership in Energy and Environmental Design," in printed materials professional discretion should be exercised.

"LEED®" – with the registered trademark superscript – is ONLY necessary for the first use in a short document, or the first use in each section of a longer document.

When a project achieves certification it should be referred to as a "LEED certified project" (not: LEED-NC certified, or LEED-CI v2.0 certified).

"LEED certification" with lowercase "c" is used to describe the certification process.

"LEED certified" with lowercase "c" is used to describe a project that has been certified.

"LEED Certified" with capital "C" is used to describe a project that has been certified to the base level: Certified.

When a project is certified, the correct wording is "project A is LEED Silver" or
"project A is LEED certified to the Silver level"
"project A is LEED Silver certified."

"LEED Green Building Rating System" is used for the first mention of the full title. If the LEED acronym is also being used for the first time, then the ® is used.

"LEED Rating System" is used as the shorthand for the formal name.

Referencing LEED in Product Literature:

Acceptable text is; "Product A contributes toward satisfying Credit N under LEED." OR "Product A complies with B requirements of Credit N under LEED."

Unacceptable text is; "Product A is LEED [certified or qualified/ compliant/ accredited/ approved]"
"LEED product", "Product A [meets/ satisfies/ fulfills/ complies with] Credit X"

Referencing LEED Registered Projects:

Accepted text is; "LEED registered project",
"LEED Registered", "LEED candidate project",
"LEED certification candidate".
"Registered with the certification goal of ____ "
"Upon completion, this building will apply to become LEED certified"
"This project is registered with the U.S. Green Building Council and intends to pursue certification under USGBC's LEED program"
"This project is registered under the LEED Green Building Rating System"

Unaccepted text is; "This project is LEED Gold Registered". Note: Projects cannot register to achieve a specific level of certification.
"This project is LEED [Qualified/ Reviewed/ Enrolled/ Verified, etc.]." [2]

About LEED

LEED® - Leadership in Energy and Environmental Design is an internationally recognized green building certification system. It provides a third-party verification that a building or community was designed and built using strategies aimed at improving performance across all the metrics that matter most; energy savings, water efficiency, CO_2 emissions reduction, improved indoor environmental quality, and stewardship of resources and sensitivity to their impacts.

Developed by the U.S. Green Building Council (USGBC), LEED gives building owners and operators the tools they need to have an immediate and measurable impact on their building's performance. The LEED rating system is market driven, voluntary and consensus based.

LEED is flexible enough to apply to all building types – commercial as well as residential. It works throughout the building lifecycle – design and construction, operations and maintenance, tenant fitout, and significant retrofit. LEED for Neighborhood Development extends the benefits of LEED beyond the building footprint into the neighborhood it serves. [3]

The LEED rating systems address the following types and scope of projects:

1) LEED for New Construction (NC)
2) LEED for Core & Shell (CS)
3) LEED for Commercial Interiors (CI)
4) LEED for Schools (SCH)
5) LEED for Healthcare
6) LEED for Retail
7) LEED for Existing Buildings Operations and Maintenance (O&M)
8) LEED for Homes
9) LEED for Neighborhood Development.

○ ***Note:***

Projects can earn multiple certification for the same Building.

Example: Once the building earns NC, it can go on to earn certification for O&M. Or once the building earns NC, SCH etc., it can go on to earn certification for CI.

When to use which rating system?

1 LEED for New Construction was designed mainly for new commercial office buildings, but it has been applied to many other building types as well.

All commercial buildings including offices, institutional buildings (libraries, museums, churches,etc.), hotels and residential buildings of 4 or more habitable stories are eligible. LEED NC addresses design and construction activities for both new buildings and major renovations of existing buildings. [7]

2 LEED for Core and shell is a market–specific application developed to serve the speculative development market, in which project teams do not control all scopes of a whole building's design and construction. This rating system can be used for projects in which the developer controls design and construction of the entire core and shell base building (e.g., mechanical, electrical, plumbing, and fire protection systems) but has no control over the design and construction of the tenant fit-out. Examples of this type of project can be a commercial office building, medical office building, retail center, warehouse and lab facility. For these projects the owner must occupy 50% or less of the building's leasable square footage. [7]

3 LEED for Commercial Interiors addresses the specifics of tenant spaces primarily in office, retail, and institutional buildings. This system is designed to work hand in hand with the LEED for Core & Shell certification system. [7]

4 LEED for School addresses design and construction activities for both new school buildings and major renovations of existing school buildings. Other projects on school campus such as administrative offices, maintenance facilities, or dormitories are eligible for either LEED for New Construction or LEED for School. [7]

5 LEED for Healthcare was developed to meet the needs of healthcare market including the following: In-patient care facilities, licensed out patient care facilities, and licensed long-term care facilities, as well as medical offices, assisted-living facilities, and medical education and research centers. [7]

6 LEED for Retail: New Construction allows for the whole building certification of freestanding retail buildings. [7]

LEED for Retail: Commercial Interiors allows tenants to certify their build-out regardless of their control over the building envelope. [7]

7 LEED for Existing Buildings Operations & Maintenance was designed to certify the sustainability of the ongoing operations of existing commercial and institutional buildings. This rating system encourages owners and operators of existing buildings to implement sustainable practices and reduce the environmental impacts of their buildings over their functional life cycles. [7]

8 LEED for Homes - Any project that participates in LEED for Homes must be defined as a "dwelling unit" by all applicable codes. The rating system focuses specifically on single-family and small multifamily homes. [7]

9 LEED for Neighborhood Development – A large variety of project sizes are eligible. However, the project must include a residential component (new or existing). [7]

What are LEED categories?

Each LEED rating system follows a similar structure, with the strategies divided into the following categories;

1) Sustainable Sites
2) Water Efficiency
3) Energy & Atmosphere
4) Materials & Resources
5) Indoor Environmental Quality
6) Innovation in Design
7) Regional Priority

LEED for Homes has two additional categories:

- Locations & Linkages
- Awareness & Education

LEED for Neighborhood Development is organized into entirely different categories:

- Smart Location and Linkage
- Neighborhood Pattern and Design
- Green Infrastructure and Buildings

SS

WE

EA

MR

IEQ

ID

RP

LEED rating and its points system?

Each category in a LEED Rating system is made up of 'prerequisites' and 'credits'. The prerequisite are compulsory strategies that must be fulfilled, where as credits are optional strategies that may be adopted. Each credit is assigned a specific number of points depending on its environmental impact and human benefits. Prerequisites do not have any points as they are compulsory.

Each LEED Rating system has 100 base points plus 10 bonus points (6 Innovation in Design points and 4 Regional Points).

LEED for Homes is an exception which currently has 125 base points and 11 Innovation in design points. But this is expected to change to a 100 point scale when updated in 2011.

Projects achieve their certification level based on the number of points earned as follows:
CERTIFIED: 40 - 49 points
SILVER: 50 - 59 points
GOLD: 60 - 79 points
PLATINUM: 80 + points.

How to manage documents?

LEED online is a primary resource for managing LEED Documentation process. With this the project team can do the following:

1) Manage project details
2) Submit all documents online
3) Upload supporting files
4) Receive the reviewer feedback
5) Earn LEED certification

The LEED online also contains embedded calculators and tables to ensure that the package submitted to GBCI is complete and accurate.

It also has several support capabilities like:

- View CIR's
- Contact customer service
- Generate project specific reports
- Consult resources like FAQ, tutorials, offline calculators and sample documents.

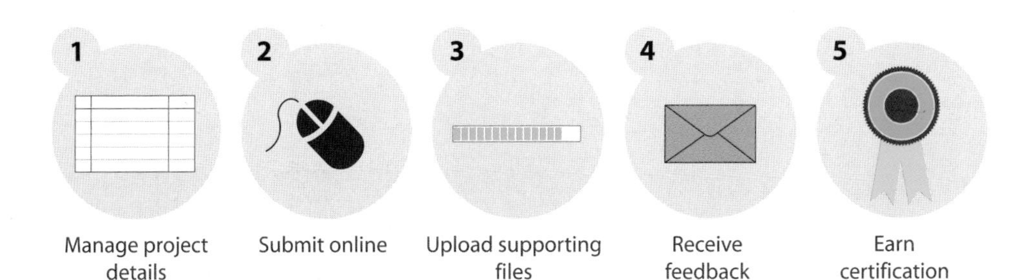

How to seek technical guidance?

In some cases, the LEED project team may face difficulties in interpreting the requirements of a prerequisite or a credit. To address this issue the Credit Interpretation Request (CIR) and ruling process was established so that project applicants could seek technical and administrative guidance. [6]

Before submitting a CIR one must:

- Review the intent of credit or prerequisite and self evaluate if the project meets the intent.
- Consult the LEED Reference Guide for detailed explanation, calculations etc.
- Review the CIR pages to see if the same inquiry has been answered before.
- If the answer is not found, contact LEED customer service to see if they can answer it.

A CIR is submitted using a online form, and following are the guidelines for its submission:

1) Do not state the credit name or your contact information.
2) Do not include confidential project details as the text will be posted on USGBC website.
3) Do not format the CIR as a letter and submit only essential background information.
4) Inquire only one credit or prerequisite at a time.
5) Do not include lengthy project narrative.
6) Submission is limited to 600 words (4000 characters including spaces).
7) Attachments are not permitted. (no plans drawings etc.)
8) Proof read the text for clarity, spellings and grammar. [6]

Following example would be the most INCORRECT way to submit a CIR:

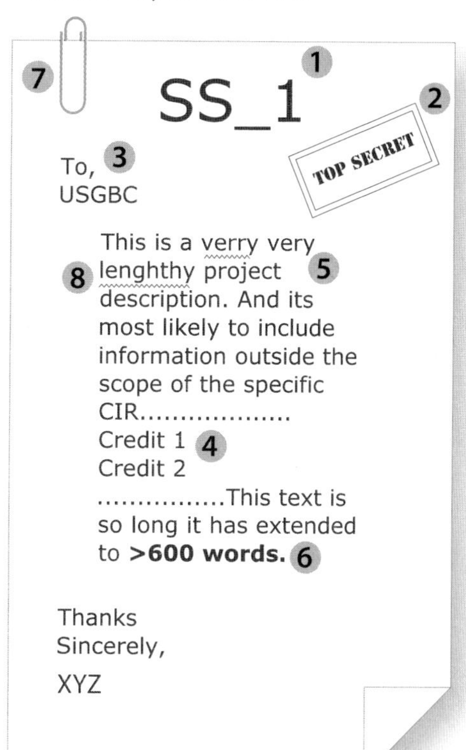

Following people have access to view CIR's:

- USGBC Company Members
- LEED Registered Project team members
- Workshop attendees for one year from the date of workshop.

How to get started?

LEED points are awarded on a 100-point scale, and credits are weighted to reflect their potential environmental impacts. Additionally, 10 bonus credits are available, four of which address regionally specific environmental issues. A project must satisfy all prerequisites and earn a minimum number of points to be certified.
The fees for certification depend on membership status and the square foot area of the building. On an average this cost is about $2000. Refer to USGBC's web site for more detail and up-to-date fees. [4]

Note/ Exceptions: *LEED for Homes and LEED for Neighborhood Development have different certification processes from the other rating systems.*

Step 1: Registration

Register projects on the GBCI web site. Registration provides access to softwares, tools and other information necessary to apply for LEED certification.

On paying the appropriate registration fee, the project will be immediately accessible in LEED Online. [4]

Step 2: Prepare Application

Each LEED credit and prerequisite has a unique set of documentation requirements that must be completed as a part of the application process.

When the necessary documentation has been assembled, the project team will upload the materials to LEED Online and start the application review process. [4]

Step 3: Submit Application

Only the LEED Project Administrator can submit an application for review. To initiate the review process, a complete application must be submitted via LEED Online.

The applications can be submitted and reviewed in either:

- One stage – Combined Design and Construction review or,
- Two stages – Split Design and Construction review.

All LEED rating systems have these two options except for LEED for Operations and Maintenance (O&M); where the applications can be submitted only in one combined stage.

Each Design and Construction review stage is further split into 2 stages:
1) Preliminary Design/ Construction review
2) Optional Final Design/ Construction review.

All project teams have the option to accept the results of the Preliminary Review as final. However, if the project team wants to submit a response to the Preliminary Review, a Final Review is conducted. [4]

○ *LEED Project boundary:*

LEED Project boundary is the portion of the building submitted for LEED certification. For multiple building developments the LEED project boundary may be a portion of development determined by the project team.

Step 4: Application Review

Upon receipt of a completed application for certification, a formal application review is initiated.

In accordance with the submittals the applications are also reviewed in either one combined stage, or two split, stages.

The Review process is conducted as follows:

- All documentation submitted are reviewed for completeness and compliance with the appropriate LEED Rating System
- Each reviewed prerequisite and credit is designated as '**anticipated**', '**pending**' or '**denied**' if it is a preliminary review, OR just '**awarded**' or '**denied**' if it's a final construction or combined review.
- All project information forms are designated as '**approved**' or '**not approved**'.
- Appropriate technical advice is given by the review team.

Appeal Review:

In the event that a project team wishes to appeal a final decision of a LEED review team, GBCI conducts an appeal review subject to the following conditions:

- All appeals of final review decisions must be filed within twenty-five (25) business days of GBCI's posting of the final review decision.
- The project team is responsible for paying the appeal review fee for each credit or prerequisite appealed ($500).
- All appeals shall be conducted in accordance with the GBCI Appeal Review Policy. [4]

Step 5: Certification

Certification is the final step in the LEED review process. Once the final application review is complete, the project team can either accept or appeal the final decision.

LEED certified projects:

1) Will receive a formal certificate of recognition
2) Will receive information on how to order plaque and certificates, photo submissions, and marketing
3) May be included (at the owner's discretion) in an online directory of registered and certified projects
4) May be included (along with photos and other documentation) in the US Department of Energy High Performance Buildings Database. [4]

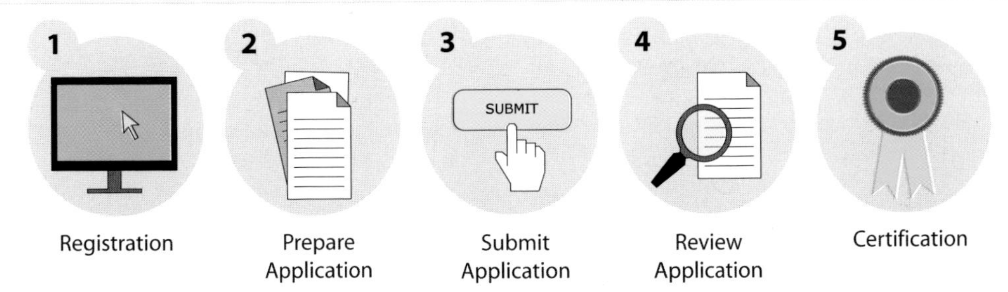

- *LEED for Core and Shell (C&S) Pre-certification, is a unique aspect of the LEED for Core and Shell program, that customers may pursue at their option. Precertification is formal recognition by USGBC given to a project for which the owner has established the goal of developing into a LEED-CS Certified building. Precertification allows C&S owners to market to potential tenants and financiers their intention to pursue unique and valuable green features in a proposed building.* [4]

Can all projects be LEED certified?

This section addresses New Construction, Core & Shell, Schools, Commercial Interiors, and Existing Buildings: Operations & Maintenance. (Excludes Homes & Neighborhood Development)

A project must adhere to the LEED Minimum Program Requirements, (MPRs) in order to achieve LEED certification.

The three goals of MRPs are as follows:

- to give clear guidance to customers,
- to protect the integrity of the LEED program,
- to reduce complications that occur during the LEED Certification process. [5]

LEED projects must comply with each applicable MPR as described below:

1 Must comply with all applicable federal, state, and local building-related environmental laws and regulations. [5]

Comply with all Laws.

2 Must be a complete, permanent building or space. Movable structures are not eligible.

- For **NC**, **C&S**, **SCH**: Must include at least one new building in its entirety.
- For **CI**: Must include a complete interior space distinct from other spaces in terms of ownership, management, lease, or party wall separation.
- For **O&M**: LEED projects must include at least one existing building in its entirety. [5]

NO movable structures;
Must be a permanent building.

3 Must use a reasonable site boundary;

a. LEED project boundary must include all contiguous land that is associated with and supports normal building operations for the LEED project building.

b. May not include land owned by somebody else unless it supports building operations for the LEED project building.

c. Any given parcel of real property may only be attributed to a single LEED project building.

d. Gerrymandering of a LEED project boundary is prohibited: the boundary may not unreasonably exclude sections of land to create boundaries in unreasonable shapes for the sole purpose of complying with prerequisites or credits.

- For **CI**: If any land was, or will be disturbed for the purpose of undertaking the LEED project, then that land must be included within the LEED project boundary. [5]

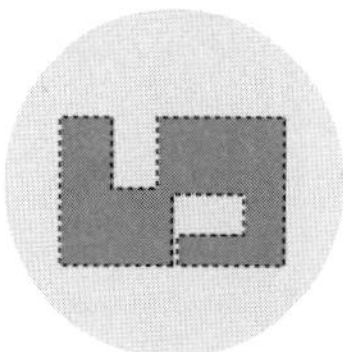

NO gerrymandering;
Reasonable Site Boundary.

4 Must comply with minimum floor area requirements. The LEED project must include a minimum of 1,000 square feet of gross floor area.

- For **CI**: The LEED project must include a minimum of 250 square feet of gross floor area. [5]

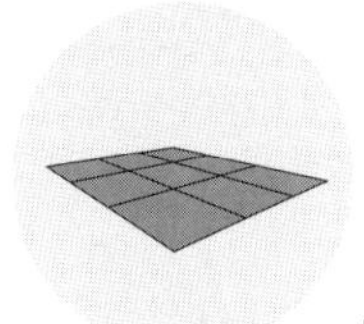

Min. floor area

5 Must comply with minimum occupancy rates.

- For **NC**, **C&S**, **SCH**, **CI**, **O&M**: The project must have minimum one (1) Full Time Equivalent (FTE) occupant.
- For **O&M**: In addition to above; building systems must be operating for a period of at least 12 months prior to submission for a review. [5]

Min. occupancy rate

6 Must commit to sharing energy and water usage data with USGBC and/or GBCI for a period of at least 5 years. [5]

Share data with USGBC/GBCI.

7 Must comply with a minimum building area to site area ratio. The gross floor area of the LEED project building must be no less than 2% of the gross land area within the LEED project boundary. [5]

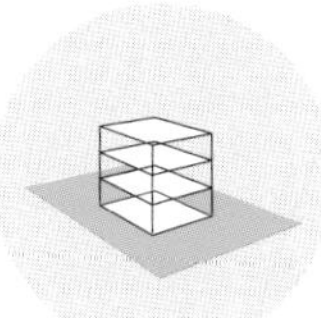

Min. building to site area ratio

○ ***Note:***

Certification may be revoked from any LEED project upon gaining knowledge of non-compliance with any applicable MPR. [5]

How to develop credit strategy?

The LEED Scorecard or LEED Credit Checklist is a list of all the credits for a specific rating system. This checklist is a very important element in the LEED integrated design process.

Next to the list of credits are 3 columns - YES, NO, MAYBE.

In a design charrette the team walks through each credit and checks the column next to it. Once completed the team knows the preliminary rating or targeted LEED Certification.

LEED 2009 for New Construction and Major Renovation

Project Checklist

Project Name: XYZ

[

Y	N	?		Sustainable Sites	Possible Points: 26
0	0	0			
Y			Prereq 1	Construction Activity Pollution Prevention	
			Credit 1	Site Selection	1
			Credit 2	Development Density and Community Connectivity	5
			Credit 3	Brownfield Redevelopment	1
			Credit 4.1	Alternative Transportation—Public Transportation Access	6
			Credit 4.2	Alternative Transportation—Bicycle Storage and Changing Rooms	1
			Credit 4.3	Alternative Transportation—Low-Emitting and Fuel-Efficient Vehicles	3
			Credit 4.4	Alternative Transportation—Parking Capacity	2
			Credit 5.1	Site Development—Protect or Restore Habitat	1
			Credit 5.2	Site Development—Maximize Open Space	1
			Credit 6.1	Stormwater Design—Quantity Control	1
			Credit 6.2	Stormwater Design—Quality Control	1
			Credit 7.1	Heat Island Effect—Non-roof	1
			Credit 7.2	Heat Island Effect—Roof	1
			Credit 8	Light Pollution Reduction	1

Y	N	?		Water Efficiency	Possible Points: 10
0	0	0			
Y			Prereq 1	Water Use Reduction—20% Reduction	
			Credit 1	Water Efficient Landscaping	2 to 4
				Reduce by 50%	2
				No Potable Water Use or Irrigation	4
			Credit 2	Innovative Wastewater Technologies	2
			Credit 3	Water Use Reduction	2 to 4
				Reduce by 30%	2
				Reduce by 35%	3
				Reduce by 40%	4

Y	N	?		Innovation and Design Process	Possible Points: 6
0	0	0			
			Credit 1.1	Innovation in Design: Specific Title	1
			Credit 1.2	Innovation in Design: Specific Title	1
			Credit 1.3	Innovation in Design: Specific Title	1
			Credit 1.4	Innovation in Design: Specific Title	1
			Credit 1.5	Innovation in Design: Specific Title	1
			Credit 2	LEED Accredited Professional	1

Y	N	?		Regional Priority Credits	Possible Points: 4
0	0	0			
			Credit 1.1	Regional Priority: Specific Credit	1
			Credit 1.2	Regional Priority: Specific Credit	1
			Credit 1.3	Regional Priority: Specific Credit	1
			Credit 1.4	Regional Priority: Specific Credit	1

Y	N	?	Total	Possible Points: 110
0	0	0		

Certified 40 to 49 points Silver 50 to 59 points Gold 60 to 79 points Platinum 80 to 110

○ ***Note:***
The credit checklist is a useful tool which can be used to develop the credit strategy and is available to non members also.

Proper site selection and its development are the first two steps towards creating sustainable buildings. Environmental damages caused by construction can take a long time to recover.

Sustainable sites are sites which addresses various measures such as:

- Selection and development of the site in wise manner
- Reducing automobile use and hence emissions associated with it
- Planting sustainable landscapes
- Protecting surrounding areas
- Managing stormwater runoff
- Reducing heat island effect
- Reducing light pollution.

This chapter explains various different strategies in detail to achieve all the above mentioned goals.

Intent:

To avoid the development of inappropriate sites and reduce the environmental impact from the location of a building on a site.

Approach:

Selecting the right site is very critical to the project. The development of site affects the ecosystem in various ways, hence it is important that the choice is made prudently.

During site selection do not include sensitive site elements and restrictive land types. The ideal sites to choose for development are, the ones which are previously developed.

LEED does NOT recommend development on following sites:

Prime farmland defined by USDA

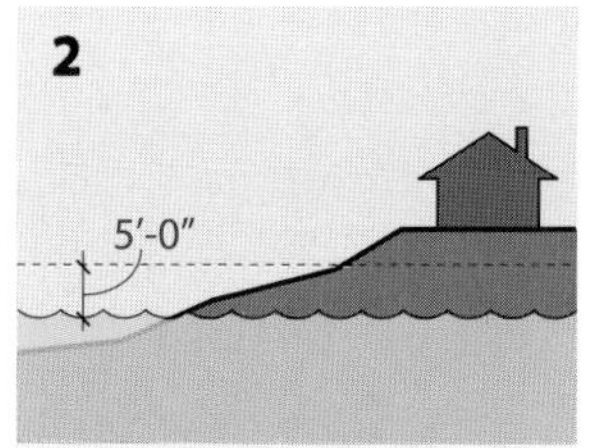

Land lower than 5 feet above 100 year flood line

Habitat for threatened or endangered species

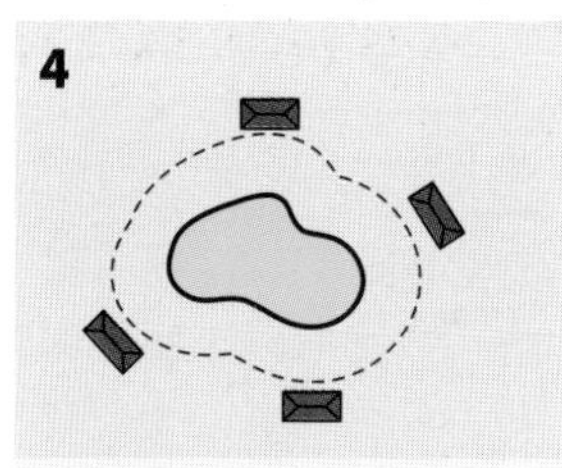

Land within 100 feet of wetlands

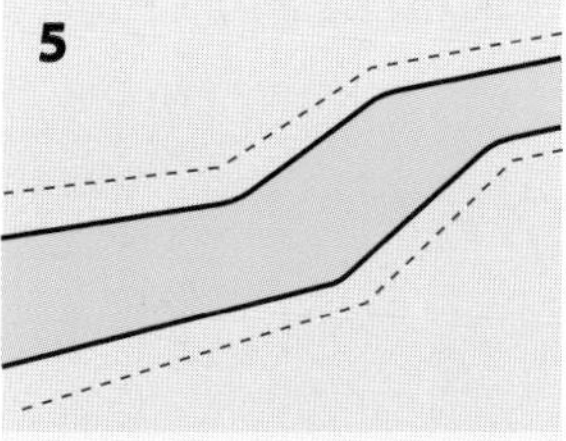

Land within 50 feet of water bodies

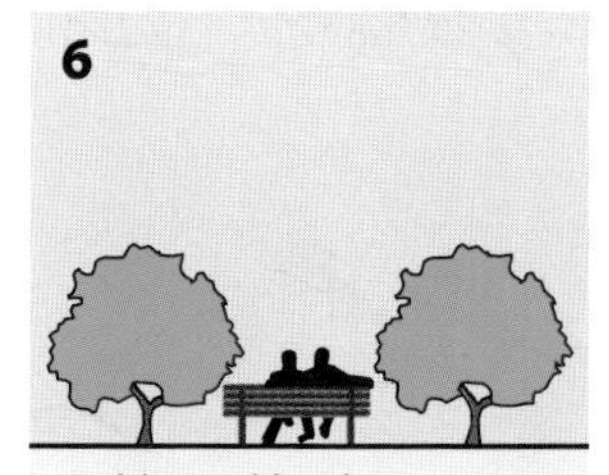

Public parkland.

Site area:

Site area is defined as total area within the project boundary, inclusive of both constructed and non constructed areas.

Intent:

To rehabilitate damaged sites where development is complicated by environmental contamination.

Approach:

EPA defines brownfield as a contaminated site with a potential presence of hazardous substances, pollutant or contaminant. [1]

Many sites in urban areas have been abandoned because of contamination from industrial activities. Such sites can be remediated and reused again. Remediation involves removing hazardous materials from sites soil and groundwater. Remediation can be expensive, but on the other side the property cost of such land can be low.

LEED encourages selection of such contaminated sites and rehabilitating them in order to reduce pressure on undeveloped land.

○ *EPA (Environmental Protection Agency):*

EPA leads the nation's environmental science, research, education and assessment efforts. The mission of the EPA is to protect human health and the environment. Since 1970, EPA has been working for a cleaner, healthier environment for the American people. [2]

Intent:

Reusing an existing building or site to protect a new undeveloped land.

Approach:

A site having pre-existing paving, construction, or altered landscape is called Previously Developed land. However, altered landscapes resulting from current agricultural use, forestry use, or use as preserved natural area are not applicable for previously developed land.

The advantage of developing on a previously developed land is having presence of already built infrastructure such as roads, utilities and services. It also helps to protect new undeveloped land.

○ ***Previously Developed Site:***
In LEED for Homes a lot consisting of 75% previously developed land is called as previously developed site. [3]

Intent:

To channel development to urban areas with existing infrastructure, protect greenfields and preserve habitat & natural resources.

Approach:

In today's time, the quality of life is greatly affected by urban sprawl.

Some negative impacts of urban sprawl are:

- Long commuting time in automobiles,
- Increased air pollution and greenhouse gas emissions.

Two factors that can help reduce urban sprawl are:

1) Development density
2) Community connectivity

LEED for Homes defines various types of densities as follows:

Density	Average housing Density
Moderate	7 Units per acre
High	10 Units per acre
Very high	20 Units per acre

LEED encourages development on sites which are within ½ mile of at least 10 basic services. (Examples of basic services are restaurants, park, banks, library, hospitals, theatres, museum, grocery stores, place of worship, school etc.) [3] People must be able to walk between the site and the services, this is called pedestrian access.

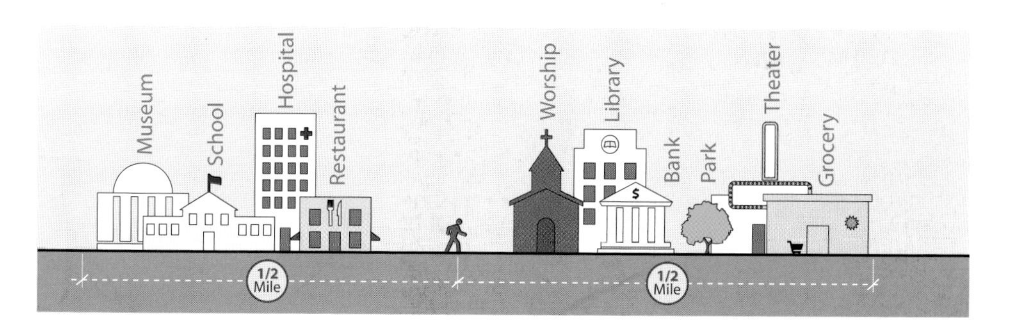

- ***FAR (Floor area ratio):***
  FAR is the measure of density of non residential land use. It is the total building floor area divided by the total buildable land area available for non-residential uses.
- ***Building Footprint:***
  Building footprint is the part of site area occupied by the building structure, excluding parking lots, landscapes etc.
- ***Density radius:***
  Density radius is calculated to determine if the project site is located in a dense area or not.
- ***Street Grid density:***
  This is an indicator of neighborhood density, calculated as the number of centerline miles per square mile.

Intent:
To reduce pollution caused by automobiles, by encouraging people to take public transport.

Approach:
The single-occupancy vehicles rely extensively on petroleum and contribute to pollution. Thus the use of mass-transit is ideal as it:

- Reduces the energy demand for transportation
- Reduces greenhouse gas emissions
- Reduces the requirement for parking lots.

The project should be located at walkable distances from commuter rail and bus stops. In certain areas where the mass transit is not in close proximity of the site, provision of shuttle service should be considered to help people reach transportation.

LEED credit can be achieved by those projects, which are located near mass transit, as it facilitates the people to take public transport and keep their cars off the roads.

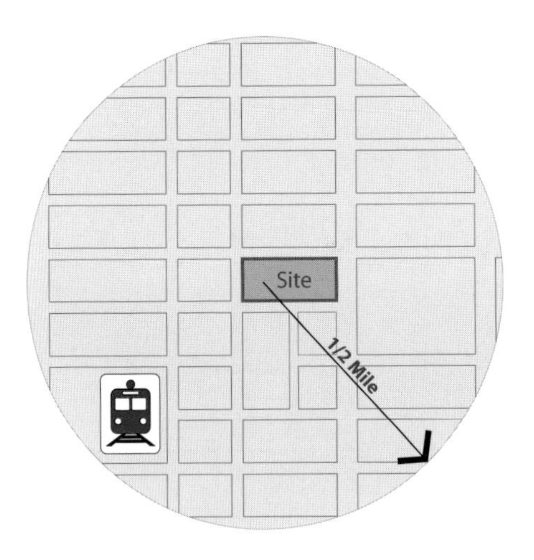

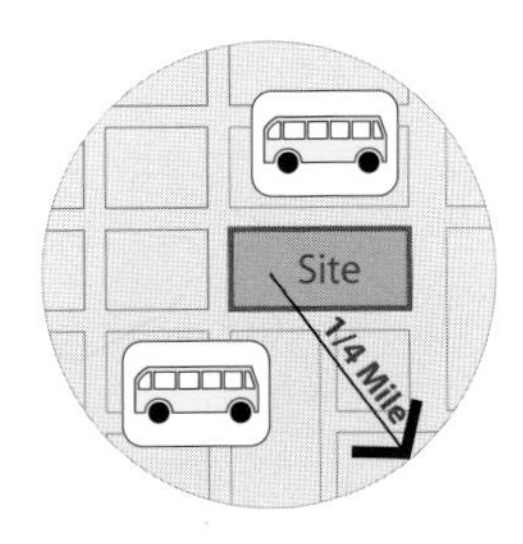

○ ***Public transportation:***
It consists of bus, rail, or other transit services for the general public that operate on a regular basis.

Alternative Transportation - Bicycle storage and changing rooms

Intent:
To reduce pollution caused by automobiles, by encouraging people to use bicycles.

Approach:
Using a bicycle as an alternative to personal vehicle has tremendous benefits. Bicycles require zero fuel, reduce noise and air pollution, they relieve traffic congestion, and require minimum infrastructure for road and parking.

The provision of bicycle racks, shower and changing facilities, in or near the building encourages the employees to bike to work.

For a commercial project LEED credit can be obtained by providing secured bicycle racks, shower and changing facilities within 200 yards of a building entrance.

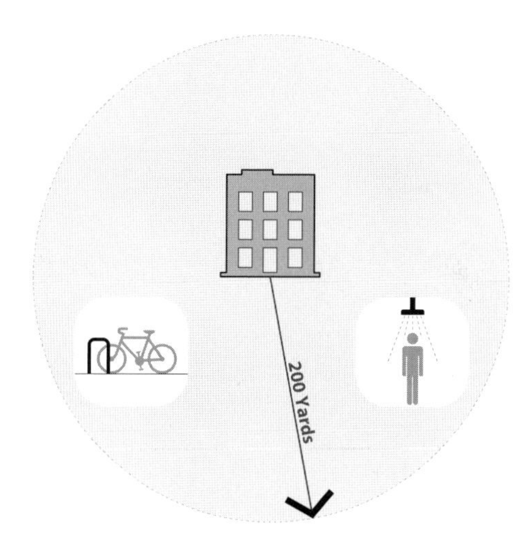

○ *Transient Users:*
Transient Users are occupants who do not use a facility on a regular, daily basis. Examples are customers in retail store, visitors in institution and college students in a classroom etc.

Intent:
To reduce pollution caused by automobiles, by encouraging people to use alternative fuel vehicles.

Approach:
Vehicles contribute to air pollution and global warming by emitting pollutants and green house gases. Alternative-fuel and fuel efficient vehicles are the promising vehicle substitutes of the future, as they produce environmental benefits.

Low Emitting vehicles (LE) are vehicles classified as Zero Emission Vehicles (ZEV) by the California Air Resources Board.

Fuel-Efficient vehicles (FE) are vehicles that have a minimum Green Score of 40 by the American Council for an Energy Efficient Economy. (ACEEE)

To encourage the use of LE/FE vehicles LEED outlines following options:

1) Providing preferred parking for LE/FE vehicles. Giving discounted parking rates to the LE/FE vehicle owners is a valid substitute for preferred parking.
2) Installing alternative-fuel fuelling station.
3) Providing LE/FE vehicles to employees.
4) Providing access to LE/FE vehicle sharing program.

1

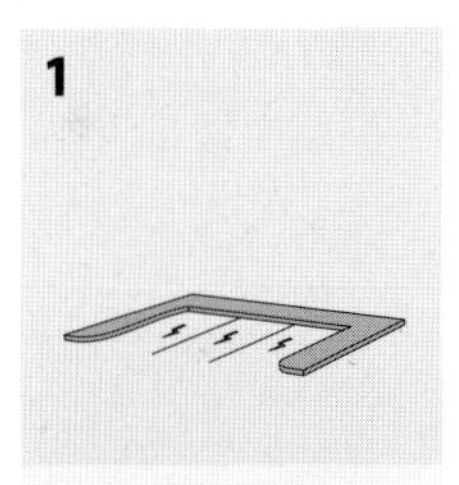

Preferred parking for LE/FE

2

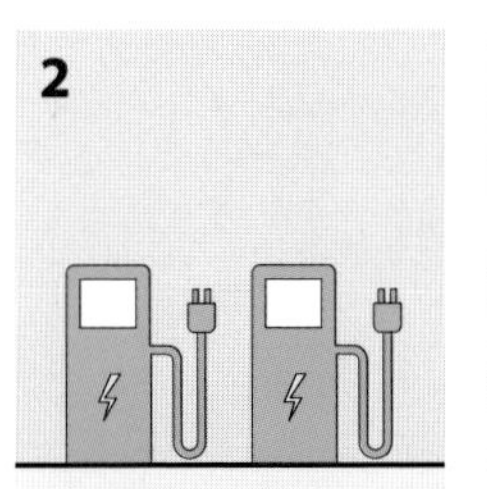

Alternative-fuel fuelling station

3

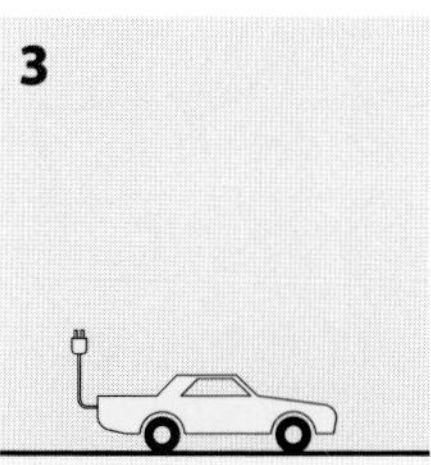

LE/FE vehicles to employees

4

LE/FE vehicle sharing program.

○ *Alternative-fuel Vehicles:*
Alternative fuel vehicles use low-polluting, non-gasoline fuels, such as electricity, hydrogen, propane or compressed natural gas, liquid natural gas, methanol and ethanol. For the purpose of LEED the gas-electric hybrid vehicles are also considered as alternative-fuel vehicles.

Intent:
To reduce pollution and land development impacts caused by automobiles, by discouraging parking spaces.

Approach:
A limited parking discourages automobile use. Reducing private automobiles saves energy and prevents pollution, as well as environmental impacts associated with oil extraction. Parking facilities also have negative impacts because of the asphalt surface, as it increases stormwater runoff and contributes to heat island effect.

Carpooling reduces the need for huge parking areas. It also helps to reduce building cost, as less land is needed for parking and infrastructure.

Preferred parking is the parking available to specific users, for example:
- designated spaces close to the building
- designated covered spaces
- discounted parking passes
- guaranteed passes in a lottery system

LEED recommends various strategies for this credit such as:
1) Having minimum parking, required by the local zoning codes and not exceeding them.
2) Providing preferred parking to carpools and vanpools.
3) Providing NO new parking.

1

P ≤ Min. Zoning Req.

Minimum parking

2

Preferred parking to carpools and vanpools

3

No new parking

- *Parking footprint:*
  Parking footprint is the area of site occupied by parking structure.
- *Car-sharing:*
  Car-sharing is a type of car renting where people can reserve cars and use them for a short period of time, often by the hours. The cars are located throughout the city and often located for access by public transport. e.g. Zip car.

Intent:

To conserve existing natural areas and restore damaged areas to provide habitat and promote biodiversity.

Approach:

Development often damages site ecology, indigenous plants and regional animal populations. Restoring natural areas on the project site benefits the environment and the society. The damaged sites can be healed by restoring native and adapted vegetation.

Native and Adapted vegetation:
These are the plants which are indigenous to a locality or plants which are adapted to local climate.

Following are the benefits of using native or adapted plants:

- They require less maintenance compared to non-native plants.
- They use minimum fertilizers, pesticides, and water.
- They reduce the costs over the building life cycle.
- They provide habitat value.

Invasive plants are non native plants and are likely to cause harm once introduced in the ecosystem. These are aggressive and are among the greatest threat to the stability of ecosystem.

○ ***Landscape area:***
Landscape area of the site is the total site area, minus building footprint, hardscape area, water bodies etc.

Intent:

To promote biodiversity by providing a high ratio of open space.

Approach:

Open spaces in urban areas have various benefits. Open spaces are a habitat for plants and animals. They reduce heat island effect and increase stormwater infiltration. These open spaces provide us a means of connection with nature.

For LEED project open space is defined as property area minus development footprint. These open areas must be vegetated and pervious.

LEED permits inclusion of following spaces in the calculation of open spaces:

- Vegetated roof areas of projects located in dense urban areas.
- Pedestrian oriented hardscapes of projects located in dense urban areas.
- Wetlands/ natural ponds with side slope of 1:4 or less.

Open spaces can be maximized in following ways:

1) Reducing the parking footprint and compacting roads.
2) Reducing building footprint by stacking the program.
3) Sharing same space for multiple usage.
4) Tuck-under parking.

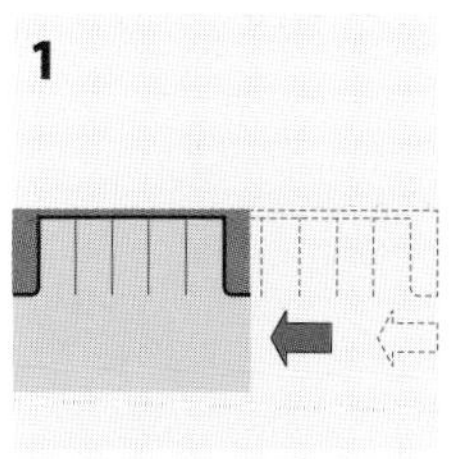

Reduce parking footprint

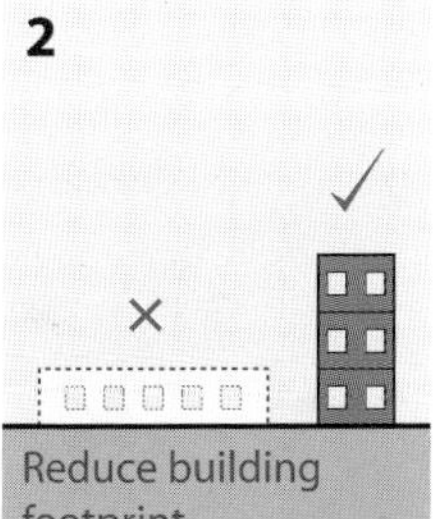

Reduce building footprint

Share space for multiple usage

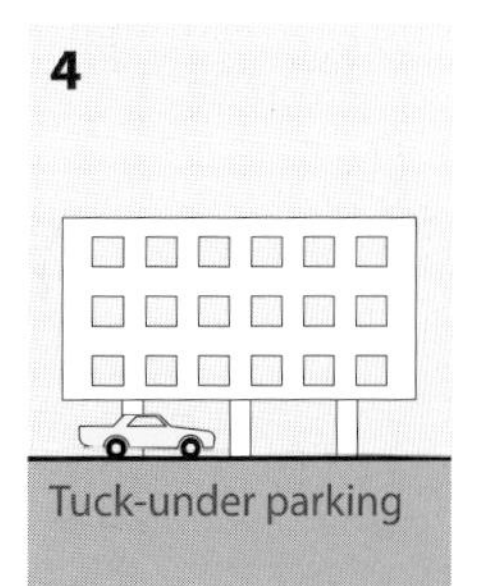

Tuck-under parking

o ***Parking footprint:***
Parking footprint is the area of site occupied by parking structures.

Intent:
To limit the disruption of natural hydrology.

Approach:
The stormwater runoff is the water from rains and melted snow that flows over surfaces into sewer systems or receiving bodies.

The increase in the quantity of stormwater runoff due to development can cause erosion, widen channels and cause downcutting in streams.

Effective stormwater management practices allow stormwater to infiltrate in the ground thereby reducing its volume.

Following are some of the ways to minimize stormwater runoff volume:

1

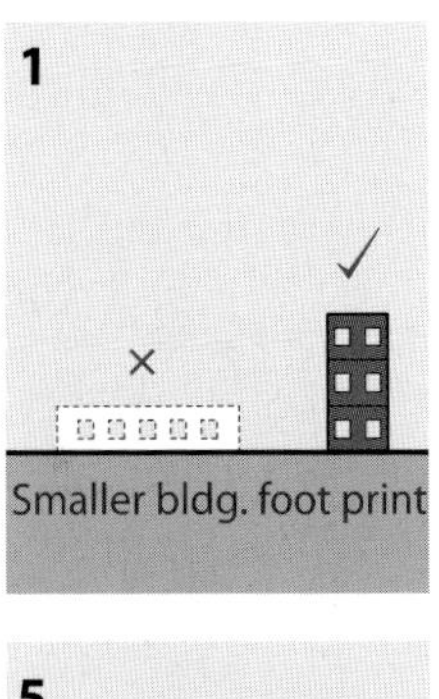

Smaller bldg. foot print

2

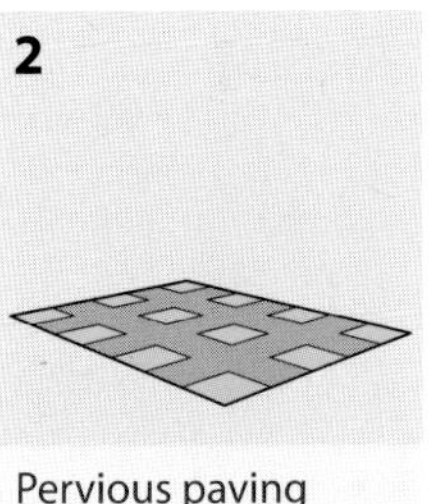

Pervious paving

3

Roof garden

4

Stormwater harvesting

5

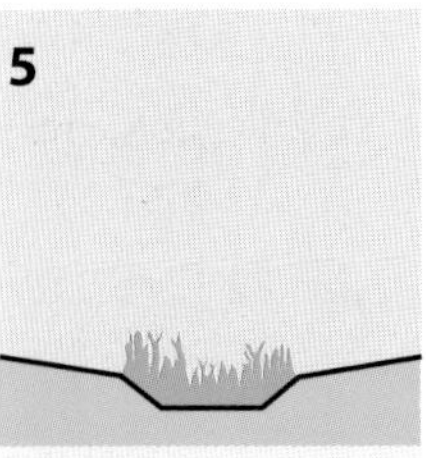

Bioswales

6

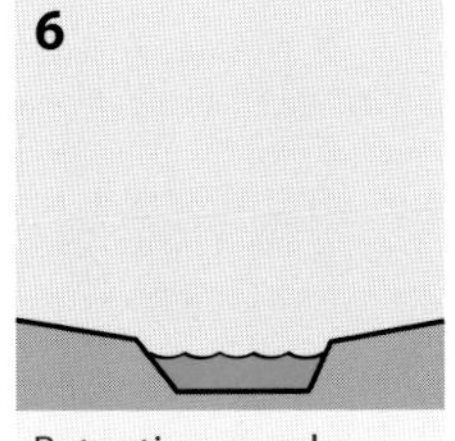

Retention pond

7

Cluster development

- ***Impervious surfaces:***
  Impervious surfaces have a perviousness of less than 50% and promote runoff of water instead of infiltration into the subsurface. Examples include parking lots, roads, sidewalks, and plazas.

Intent:
To limit the disruption of natural hydrology.

Approach:
In developed urban areas the permeability of surfaces is greatly reduced. This causes the stormwater runoff to flow into gutters and sewers and finally into the receiving waters. This stormwater contains sediments and other contaminants which deteriorate the water quality. Hence, it is necessary to reduce the impervious areas.

Following are two different storm water management measures, which can be used to minimize impervious areas:

Non-structural measures: The stormwater naturally percolates into the soil, gets filtered and most pollutants are broken down by micro-organisms.
1) Rain gardens
2) Vegetated swales
3) Disconnected impervious areas and pervious pavements

Structural measures:
1) Rainwater cistern
2) Manhole treatment device
3) Ponds.

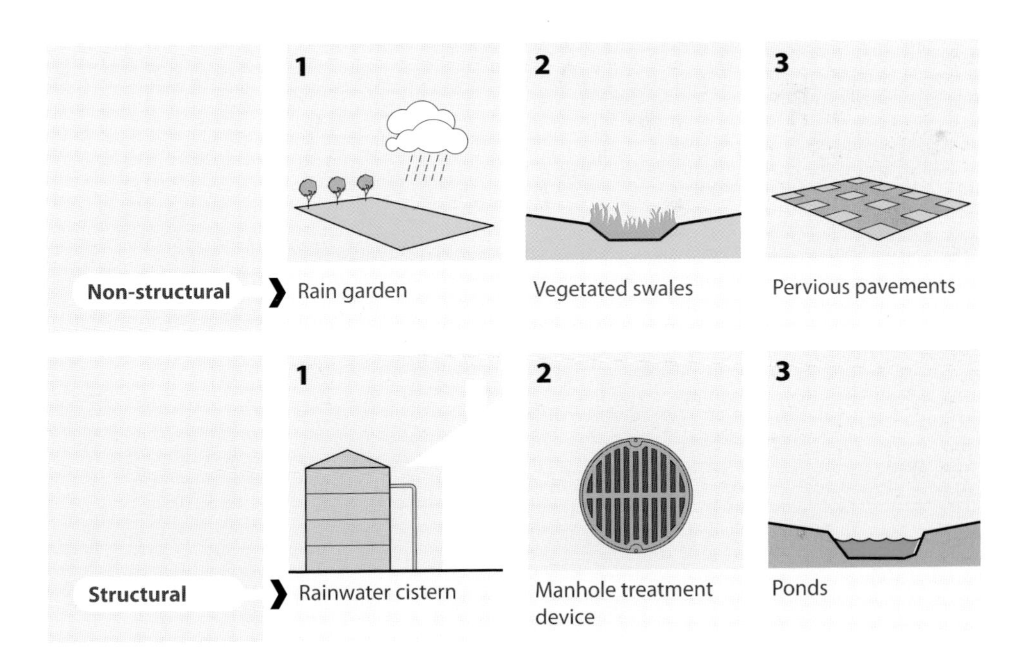

○ ***Chemical Runoff:***
When the rainwater falls on parking lots, streets and landscapes, it gets contaminated and contains oil, pollutants and fertilizers, this water is called chemical runoff.

Intent:
To reduce heat islands caused from non- roof areas.

Approach:
Heat island effect refers to the absorption of heat by hardscapes, such as dark, non-reflective pavement and buildings, and its radiation to surrounding areas. Particularly in urban areas, other sources may include vehicle exhaust, air conditioners, and street equipment; reduced airflow from tall buildings and narrow streets exacerbate the effect.

Due to this effect the ambient temperature in urban areas is about 10°F more than the surrounding undeveloped areas.

LEED recommends following strategies for reducing the heat island effect:
1) Providing shade from existing trees, or within 5 years of landscape installation.
2) Providing shade with the help of structures covered with solar panels.
3) Providing shade with architectural devices having a SRI of at least 29.
4) Using hardscape materials with minimum SRI of 29.
5) Using open grid pavement system (at least 50% pervious)
6) Placing 50% parking under cover.

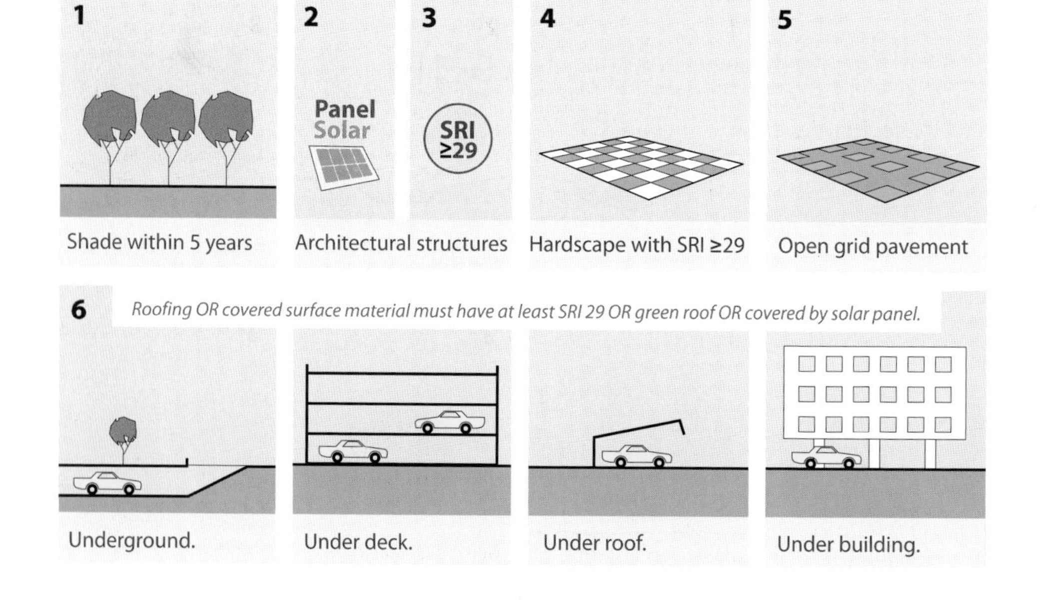

- ***Open grid pavement:***
  Open grid pavement is less than 50% impervious and has vegetation in the open cells.
- ***Solar Reflective Index (SRI):***
  Solar Reflective Index is a measure of materials ability to reject solar heat. SRI is calculated using emittance and reflectance value. SRI of standard black surface is 0 and SRI of standard white surface is 100. Materials with the highest SRI are the best choice for the paving.

Intent:

To reduce heat islands caused from roof areas.

Approach:

The roof area comprises of a substantial portion of the building envelope. The roof also gets the radiation from the sun for the most time of the day and hence it is a major factor in contributing to the heat island effect. The energy used to cool a building represents a substantial portion of the building operating budget.

The heat island caused due to roofs can be minimized in following ways:

1) Using High-Reflectance materials on the roof. (For LEED, minimum SRI of 78 for low sloped roof and a minimum SRI of 29 for a steep-sloped roof is required).
2) Installing a vegetated roof (also know as green roof)
3) Doing a combination of above two strategies.

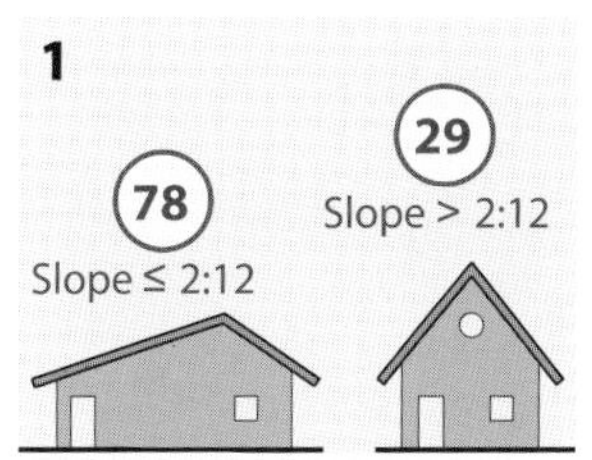

High reflectance material on roof

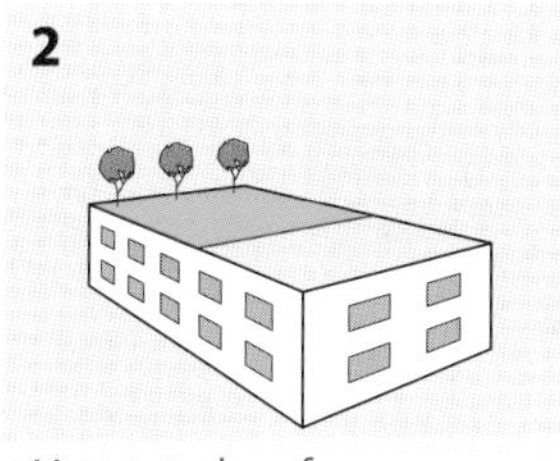

Vegetated roof

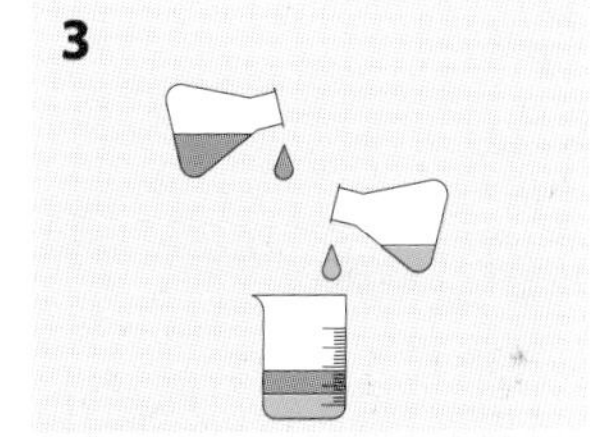

High reflectance material on roof + Vegetated roof

○ ***Vegetated roof/green roof:***

This is a system in which vegetation is grown on top of a conventional roof. This system has layers of growing medium, filter fabric, drainage and waterproof membrane on the roof.

The green roofs generally last 2-3 times longer than the conventional roof. They also retain stormwater and provide insulation benefits. The green roofs should use native plants which can be included in calucaltion of landscape areas.

Intent:

To minimize the light trespass from the buildings, to improve night time visibility and reduce impact on nocturnal environments.

Approach:

Outdoor lighting and lighting of sidewalks, pathways, parking lots, and roads is necessary for public safety and nighttime use. However, a poorly designed outdoor lighting can trespass from the site and disturb nocturnal animal life and ecosystem. The light pollution also limits the night sky observation.

LEED recommends following strategies to tackle those issues:

1) Turn off all non-emergency lights during night time or at least reduce the power input of lights.
2) All windows and openings must have automatic shielding during night (11:00 p.m. to 5:00 a.m.). This will prevent the inside light from trespassing out during the night time.
3) For exterior lights, light areas only necessary for safety and comfort.
4) Make sure that lighting densities do not exceed than those mentioned in standards for a particular zone.

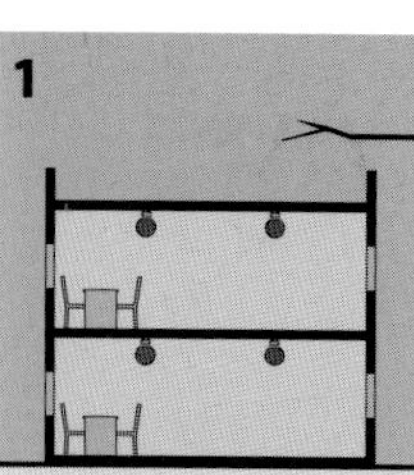

Turn off lights during night time

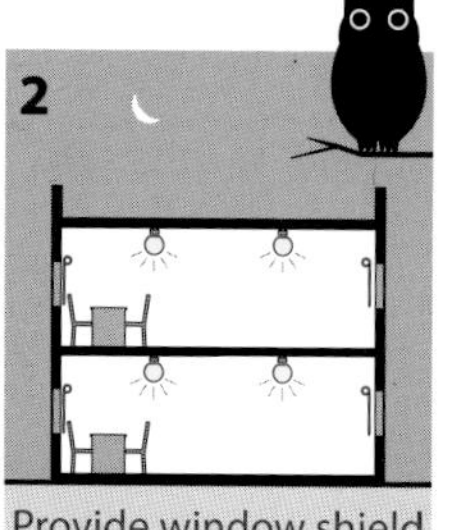

Provide window shield during night time

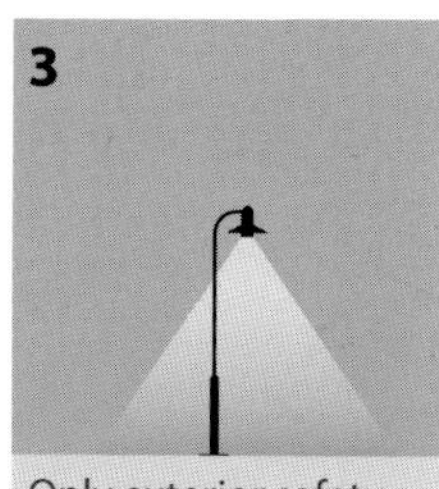

Only exterior safety lights to be on

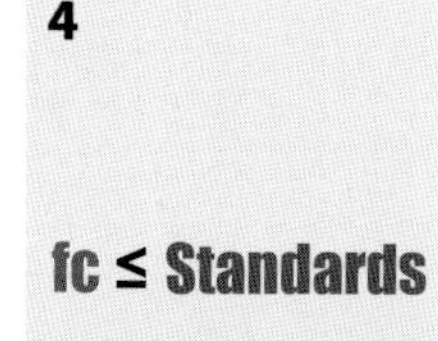

Lighting density as per standards

o ***Footcandle:***

Footcandle (fc) is a measurement of light falling on a surface. One footcandle is the quantity of light falling on a surface of one square foot area from a candle which is one foot away.

Intent:

To reduce pollution caused due to construction activities.

Approach:

During construction the most valuable top-soil from the site is lost. This top-soil is biologically active and contains plant nutrients. The loss of this soil can lead to environmental concerns, like increased use of fertilizers, pesticides, as well as increase in the quantity of storm water runoff. Hence it is very essential that this top-soil is preserved.

Before construction, it is necessary to create and implement an erosion and sedimentation control plan for all construction activities associated with the project.

The plan must describe measures implemented to accomplish the following objectives:

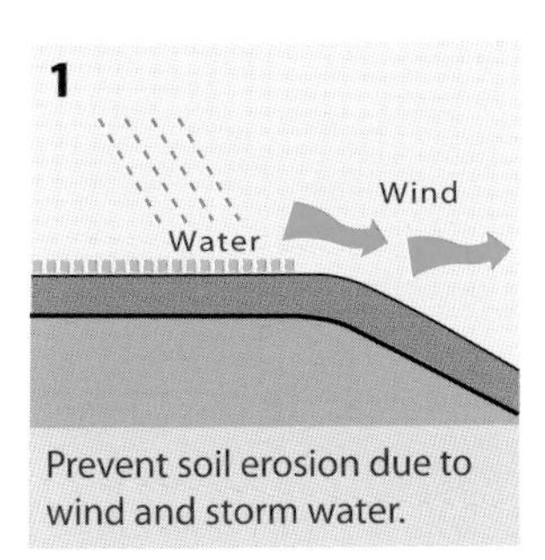

Prevent soil erosion due to wind and storm water.

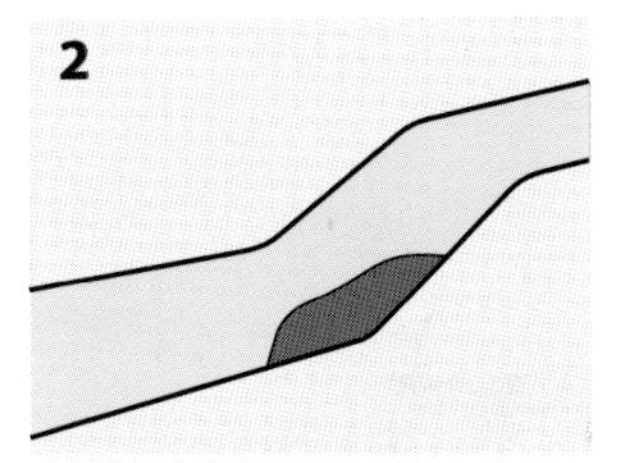

Prevent sedimentation of storm sewer or streams.

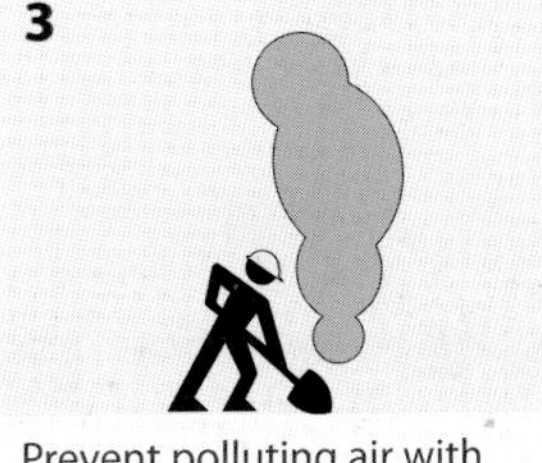

Prevent polluting air with dust.

Following are some strategies for controlling erosion and sedimentation:

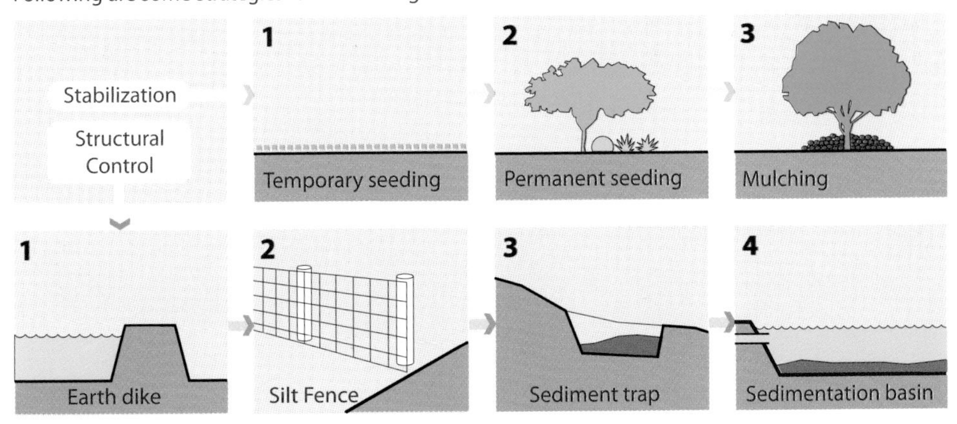

- ***Stormwater runoff :***
  It consists of water from precipitation that flows over surface into sewer systems or receiving water bodies. All water that leaves project boundary on the surface is considered as stormwater runoff.

In America the consumption of public water supply continues to increase. This limited water supply is used for domestic, commercial, industrial and other purposes. The wastewater generated, overwhelms the treatment facilities and the untreated overflow contaminates river, lakes and other potable water sources with nitrogen, bacteria, toxic metals and other contaminants.

Using large amount of water increases the life-cycle cost of a building. As per USGBC an average commercial building can achieve 30% water saving by using water efficient measures. [1]

Water Efficiency addresses environmental concerns relating to building water use and disposal.
The LEED credits aim at:

- Practicing water efficient landscaping
- Reducing the amount of wastewater generated.
- Reducing water consumption to save energy and improve environmental well-being.
- Reducing use of process water.

This chapter explains various strategies to achieve all the above mentioned intents.

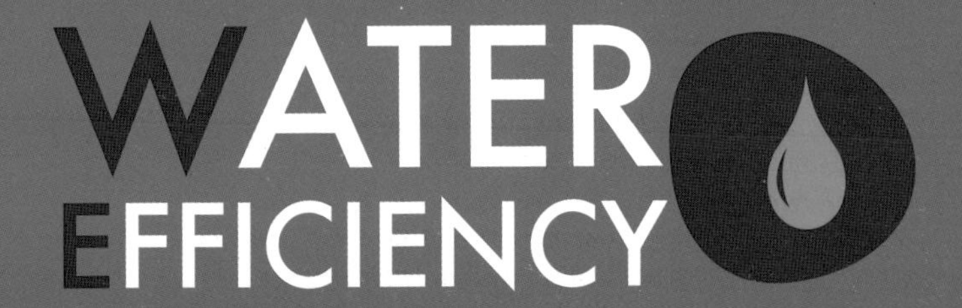

Intent:

To limit or eliminate the use of potable water for irrigation.

Approach:

In USA landscaping consumes 30% of the potable water. Increasing the landscaping efficiency can dramatically reduce or eliminate the need for irrigation. [2]

LEED specifies following methods to reduce potable water consumption for irrigation;
1) Using Plant species with low water need
2) Irrigation efficiency using micro-irrigation systems, moisture sensors, rain shut-offs, drip irrigation, and scheduling
3) Using captured rainwater
4) Using recycled waste water
5) Using water treated and conveyed by public agency specifically for non-potable use
6) Using Recycled graywater
7) Xeriscaping.

1

Plant species

2

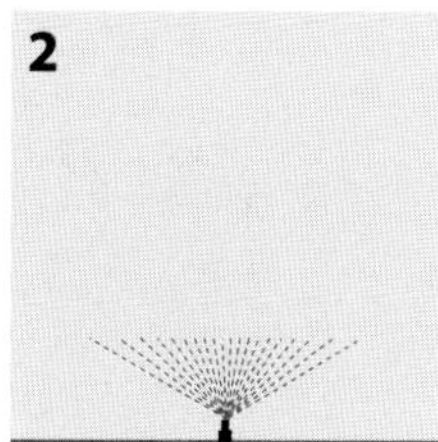

Micro-irrigation

3

Captured rain water

4

Recycled waste water

5

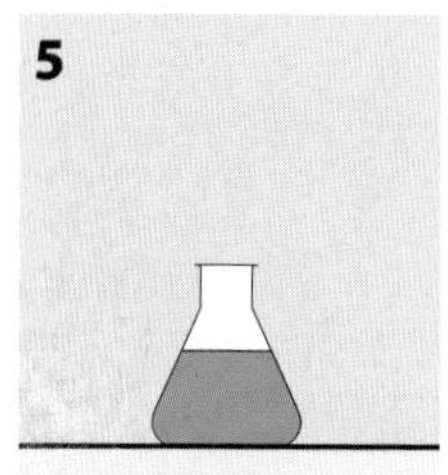

Treated water

6

Recycled gray water

7

Xeriscaping

- ***Irrigation Efficiency:***
  The percentage of water delivered by irrigation equipment that is actually used for irrigation and does not evaporate, blow away or fall on hardscape.
- ***Xeriscaping:***
  Xeriscaping is a method in which drought resistant and low water plants are used so as to conserve water, and eliminate the need for routine watering.
- ***Potable water:***
  The water which is suitable for drinking and meets EPA drinking water standards is called potable water. This water is supplied from wells and municipal water supply.

Intent:
To reduce the generation of wastewater and the demand for potable water.

Approach:
Water closets and urinals do not require drinking water for flushing. Reducing the amount of potable water for sewage conveyance reduces the amount of water drawn out of natural water bodies.

LEED recommends following strategies for waste water reduction:
1) Using water-conserving fixtures (e.g. dual flush toilets and waterless urinals)
2) Using non-potable water (e.g. captured rainwater, recycled graywater & on site or municipally treated water)
3) Treat wastewater on-site to tertiary standards. Treated water must be infiltrated or reused on-site.

Water conserving fixtures

Using non-potable water

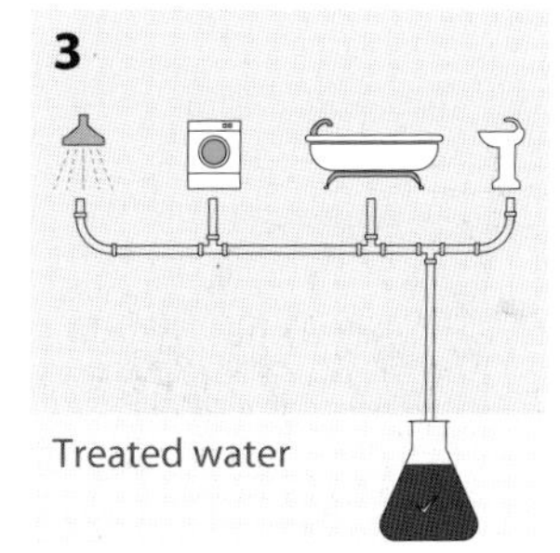

Treated water

- ***Graywater:***
  Graywater is the water that has not come in contact with toilet waste. It is the water that comes from bathtubs, showers, bathroom wash basin, washing machine and laundry tubs. It must not include water from kitchen sinks or dishwashers. However, some states and cities allow kitchen sink water to be included. Gray water is most preferred to be used for flushing toilets and urinals.
- ***Black water:***
  This is the water from toilets and urinals. However, the water from kitchen sink, showers or bathtubs is also included in the definition of black water, in some states or cities. This water in not suitable for irrigation or flushing purposes.

Intent:

To increase the water efficiency within buildings and reduce burden on municipal supply.

Approach:

It is essential to reduce the amount of potable water used in toilets, urinals, shower heads and faucets. By doing this we can preserve the water in rivers, streams and other water bodies. Reduction in water use decreases the building operation cost and brings great economic benefits.

For LEED certification, strategies should be employed, which reduce the water usage by a certain percentage, from the calculated water use baseline. *

Some strategies to reduce water use are:

1) Installing flow restrictors and/or flow aerators on sinks and showers.
2) Installing automatic faucet sensors and metering controls.
3) Installing High-efficiency toilets.
4) Installing waterless urinals
5) Using Watersense fixtures
6) Collecting rainwater.

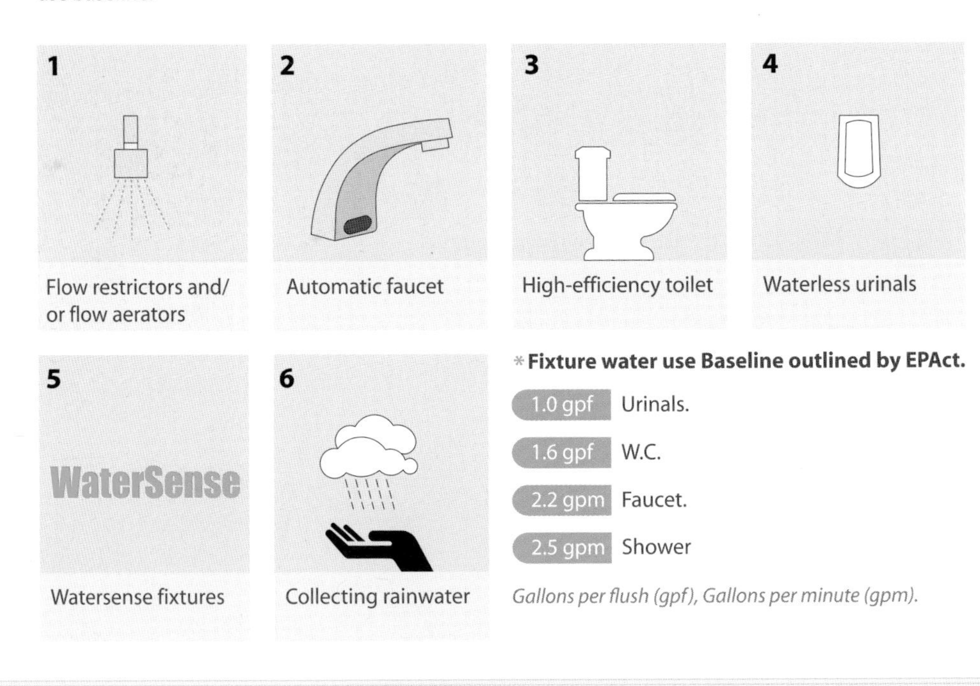

*** Fixture water use Baseline outlined by EPAct.**

1.0 gpf	Urinals.
1.6 gpf	W.C.
2.2 gpm	Faucet.
2.5 gpm	Shower

Gallons per flush (gpf), Gallons per minute (gpm).

- ***EPAct:***
  EPAct (Energy Policy Act of 1992) - A part of this act establishes water conservation standards for water closets, shower heads and faucets. This saves US approximately 6.5 billion gallons of water per day. [3]
- ***U.S. EPA, Watersense:***
  This is a partnership program sponsored by the U.S. Environmental Protection Agency, which makes it easy for Americans to save water and protect the environment. One should look for the Watersense label to choose quality, water-efficient products. [4]

Intent:

To increase the water efficiency within buildings and reduce burden on municipal supply.

Approach:

By reducing the amount of potable water used in building for process water, reduces the impact on rivers, streams and other water bodies.

When garbage disposals are used, a lot of unprocessed food is introduced in municipal water system, which in turn causes unnecessary burden on sewage treatment plants. Thus eliminating garbage disposals, preserves our resources and saves energy. Food waste can be instead turned into compost, which is environmentally beneficial and enriches the soil.

LEED for School recommends 20% reduction of water use from the baseline in following, process items to earn a credit:

1) Clothes washer
2) Dishwashers
3) Ice machines
4) Food steamers
5) Prerinse spray valves.

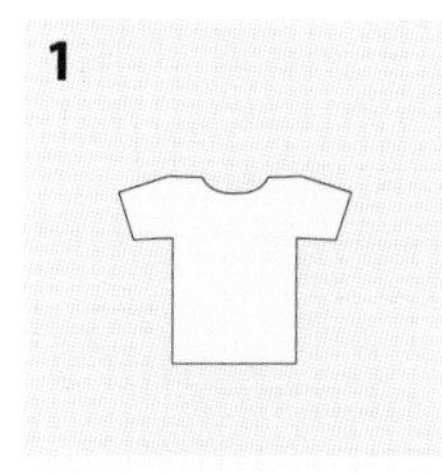

1 Clothes washer

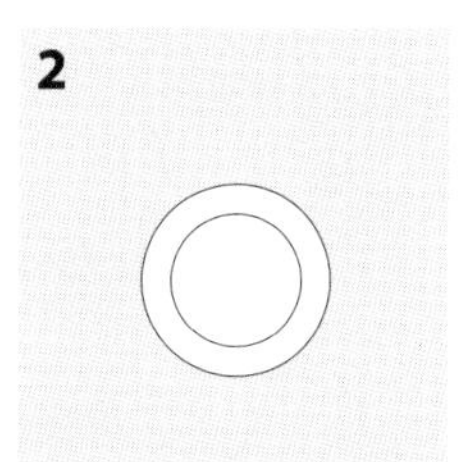

2 Dishwasher

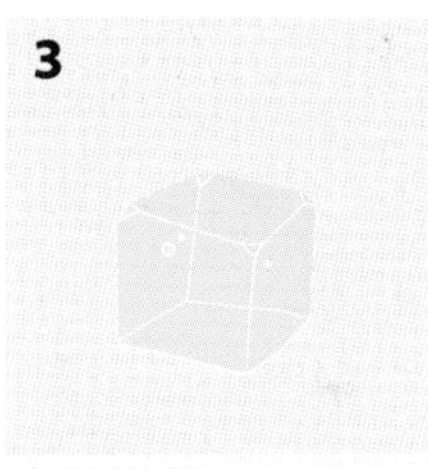

3 Ice machine

4 Food steamers

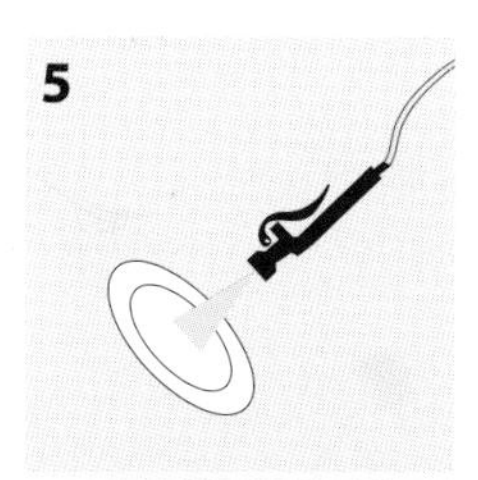

5 Prerinse spray valves

○ ***Process water:***
Process water is the water used in dishwashers, clothes washer, ice machines, cooling towers, boilers, chillers and other industrial equipments. This water is not regulated by Energy Policy Act of 1992. [3]

According to the U.S. Department of Energy, buildings consume approximately 39% of the energy and 74% of the electricity produced annually in the United States.[1] The electricity generated from fossil fuels, such as oil, natural gas, and coal, has negative effects on the environment at every stage of its production and use.

Electricity is generated by burning fossil fuels. Burning coal releases harmful pollutants, such as carbon dioxide, sulphur dioxide, nitrogen dioxide, small particulates and mercury.

Natural gas, nuclear power plants and hydroelectric generators have adverse environmental effects as well. Natural gas is a source of nitrogen oxide and greenhouse gas emissions. Nuclear power plant has safety and waste disposal issues. Hydroelectric plants disrupt natural water flow and their habitats.

These issues are addressed by green buildings in two ways
- They reduce the amount of energy required for building operation.
- They make use of more eco-friendly forms of energy.

The Energy and Atmosphere category has the most points and the greatest weightage in all the different LEED rating systems. Following are the various topics covered by this category.
- Energy Performance
- Tracking Building Energy performance
- Managing Refrigerants to eliminate CFCs
- Using Renewable Energy.

This chapter elaborates the above topics in detail.

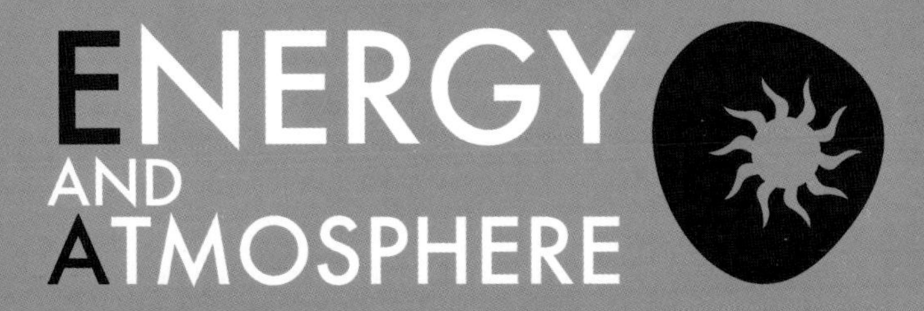

Intent:

To achieve increasing levels of energy performance in order to reduce environmental and economic impacts associated with excessive energy use.

Approach:

Fossil fuels such as coal and oil, are the most common sources of energy used in building. However, these fuels are finite and have many environmental impacts, including air and water pollution, land degradation, waste generation and greenhouse gas emission. It is very important that our buildings are energy efficient, in order to reduce the environmental burden associated with energy production and consumption.

There are 4 fundamental strategies which can increase energy performance:

1) Reducing energy demand by optimizing building form and orientation, reducing internal loads through lighting improve ments, and shifting load to off-peak periods,
2) Harvesting free energy by using onsite resources, such as daylight, ventilation-cooling, solar heating & power, and wind energy for space conditioning.
3) Increasing efficiency with more energy efficient building envelope, lighting system, and HVAC system.
4) Recovering waste energy through exhaust air energy recovery systems, graywater heat recovery systems and cogeneration.

LEED points for this credits can be achieved in following ways:

- **Whole Building Energy Simulation** – Demonstrating a percentage improvement in the proposed building performance rating using computer simulation model.

- **Prescriptive compliance path: ASHRAE Advanced building Design** – Comply with its prescriptive measures.

- **Prescriptive compliance path: Advanced building™ Core Performance™ Guide** – Comply with its prescriptive measures.

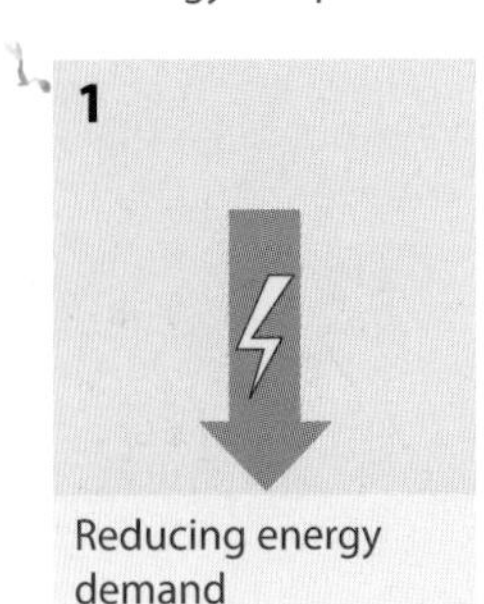

Reducing energy demand

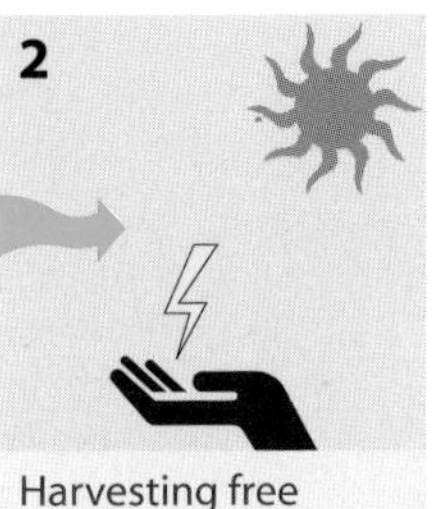

Harvesting free energy

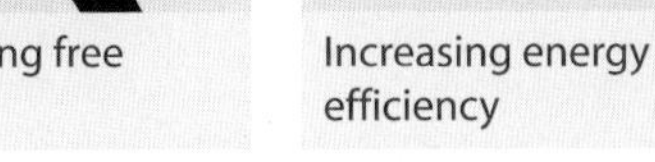

Increasing energy efficiency

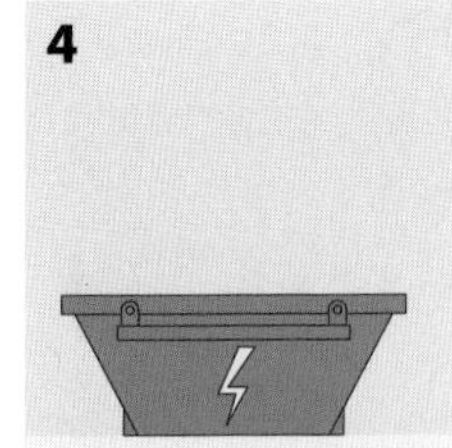

Recovering waste energy

Proposed building performance :

Proposed building performance is the annual energy cost calculated for a proposed design as defined in ASHRAE 90.1-2007, Appendix G.

ASHRAE:

ASHRAE - American Society of Heating, Refrigerating, and Air-Conditioning Engineers (pronounced as "ash-ray") is an international organization with a mission of advancing the arts and science of heating, ventilation, air conditioning and refrigeration to serve humanity and promote a sustainable world through research, standards writing, publishing and continuing education. [2]

Following are the ASHRAE standards which are used and referred in LEED V3:

- ASHRAE standard **55- 2004** – Thermal comfort conditions for human occupancy.
- ASHRAE standard **62.1- 2007** – Ventilation for acceptable indoor air quality.
- ASHRAE standard **90.1- 2007** - Establishes minimum requirements for the energy efficient design of the building.
  Components of ASHRAE 90.1-2007 are as following:
  Section 5 – Building Envelope
  Section 6 – Heating, Ventilation and Air conditioning
  Section 7 – Service water heating
  Section 8 – Power
  Section 9 – Lighting
  Section 10 – Other Equipments. [2]

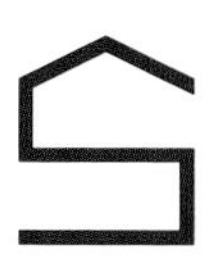
Section 5

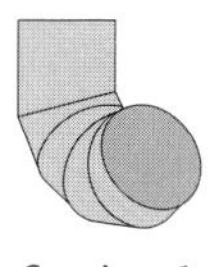
Section 6

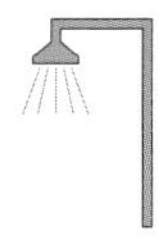

Section 7

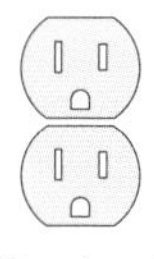
Section 8

Section 9

Section 10

ICC:

The International Code Council develops building safety and fire prevention codes used to construct residential and commercial buildings, including homes and schools. [3]

ADA:

ADA - Americans with Disabilities Act, document sets guidelines for accessibility to places of public accommodation and commercial facilities by individuals with disabilities. These guidelines are to be applied during the design, construction, and alteration of such buildings and facilities to the extent required by regulations issued by Federal agencies, including the Department of Justice, under the Americans with Disabilities Act of 1990. [4]

ISO 14000:

The ISO 14000 is a standard for environmental management systems that is applicable to any business, regardless of size, location or income. The aim of the standard is to reduce the environmental footprint of a business and to decrease the pollution and waste a business produces. The most recent version of ISO 14001 was released in 2004 by the International Organization for Standardization (ISO) which has representation from committees all over the world. [5]

Intent:
To encourage the use of on-site renewable energy.

Approach:
In US, electricity is generated primarily from combustion of fossil fuels, large hydroelectric dams, or nuclear power plants. These traditional approaches present unique environmental concerns.

Renewable energy is derived from sun, wind, water, or the Earth's core. It can also be derived from biomass or plant matter. This renewable energy has lower pollution emissions, reduced health risks, and helps preserve natural resources. Every kiloWatt hour (kWh) of renewable power avoids one pound of CO_2 emission.

LEED recommends using renewable energy systems to offset building energy costs.

Following are the on-site renewable energy systems eligible for the LEED credit:
1) Photovoltaic systems
2) Wind energy systems
3) Solar thermal systems
4) Biofuel-based electrical systems
5) Geothermal heating systems
6) Geothermal electric systems
7) Low-impact hydroelectric power systems
8) Wave and tidal power systems

Following biofuels are considered renewable energy for the purpose of LEED credit:
- Untreated wood waste, including mill residue.
- Agricultural crops or waste.
- Animal waste and other organic waste.
- Landfill gas.

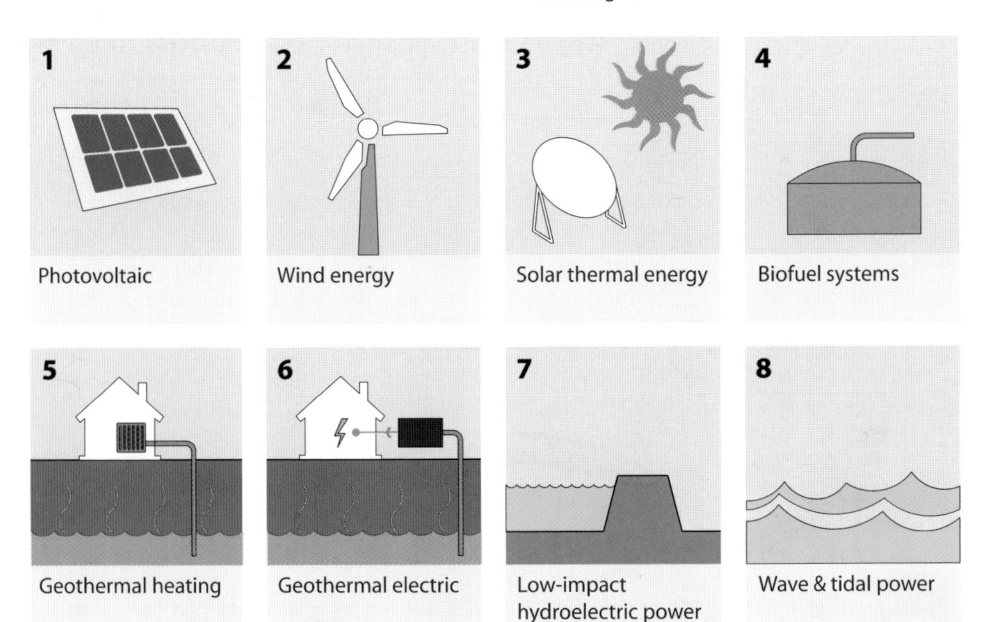

○ ***Biomass:***
Biomass is plant material from trees, grasses, or crops that can be converted to heat energy to produce electricity.

Intent:

To verify that the projects energy related systems are installed, calibrated and perform as required.

Approach:

Commissioning has various benefits such as:

- lower operation costs,
- fewer contractor callbacks,
- better building documentation,
- improved occupant productivity and
- verification that the systems perform in accordance with the owners project requirements (OPR).

LEED mandates certain commissioning process activities and gives credits for other additional commissioning process activities.

LEED requires that commissioning process activities must be completed for the following energy related systems, at a minimum:
1) HVAC & Refrigeration systems
2) Lighting and Daylight controls
3) Domestic hot water systems
4) Renewable energy systems (wind, solar, etc.)

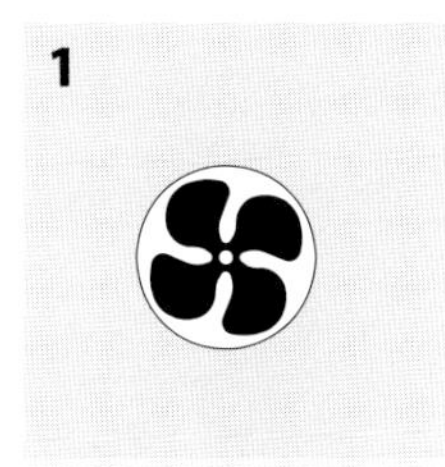

HVAC & R systems

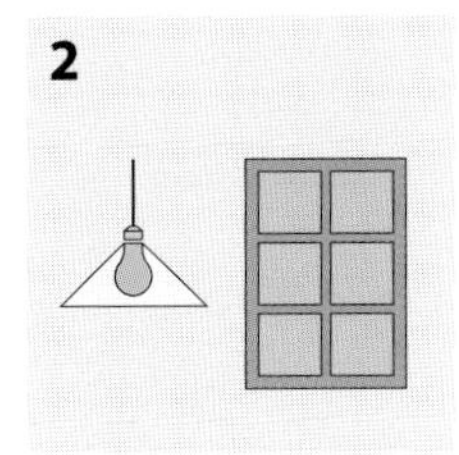

Lighting & daylighting controls

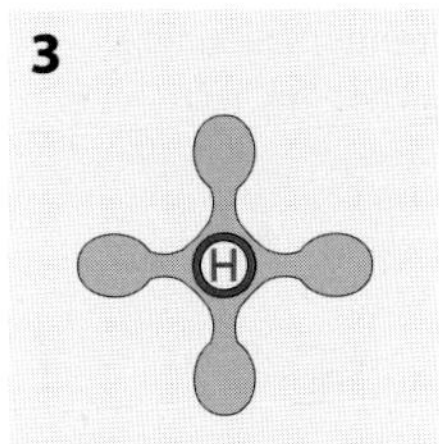

Domestic hot water systems

Renewable energy systems

○ ***Owner's project requirements (OPR):***
Owner's project requirements is a written document that details the ideas, concepts, and criteria that are determined by the owner to be important to the success of the project.

Intent:

To encourage the use of grid-source renewable energy.

Approach:

When energy is generated from traditional sources like coal, natural gas and other fossil fuels; air pollutants such as sulfur dioxide, nitrogen oxide and carbon dioxide are released. These air pollutants are the primary cause of smog, acid rain and climate change.

Green electricity relies on renewable energy sources, such as solar, water, wind, biomass and geothermal source, and thus reduces the air pollution caused due to electricity generation.

LEED recommends the use of electricity from renewable sources. Buildings can obtain a LEED credit if they engage in a 2 year renewable contract to provide at least 35% of the building's energy from renewable sources, as defined by the Center for Resource Solutions' Green-E Energy product certification.

There are 3 approaches to achieve this LEED credit:

1) In a state with open electricity market, select Green-e -certified power provider for a 2 year contract period.
2) In a state with closed electricity market, the utility company may have a Green-e-accredited program. In this case, enroll the building in renewable power program for 35% of provided electrical energy.
3) If Green-e power cannot be purchased, the team can purchase renewable energy certificates (RECs).

o ***Geothermal energy:***

Geothermal energy is electricity generated by converting hot water or steam from within the earth into electrical power.

What is Green-e?

Green-e Energy is the nation's leading independent certification and verification program for renewable energy. It is a voluntary consumer-protection program that certifies superior renewable energy options offered by utilities and marketers. It is administered by Center for Resource Solutions, a nonprofit based in San Francisco, CA.

The following three types of renewable energy options are eligible for Green-e Energy certification:

- Renewable Energy Certificates (REC's; also known as 'green tags')
- Utility Green Pricing Programs
- Competitive Electricity Products

REC's

When a renewable energy facility operates, it creates electricity that is delivered into a vast network of transmission wires, often referred to as "the grid." The grid is segmented into regional power networks called pools. To help facilitate the sale of renewable electricity nationally, a system was established that separates renewable electricity generation into two parts: the electricity or electrical energy produced by a renewable generator and the renewable "attributes" of that generation. (These attributes include the tons of greenhouse gas that were avoided by generating electricity from renewable resources instead of conventional fuels, such as coal, nuclear, oil, or gas.) These renewable attributes are sold separately as renewable energy certificates (RECs). One REC is issued for each megawatt-hour (MWh) unit of renewable electricity produced. The electricity that was split from the REC is no longer considered "renewable" and cannot be counted as renewable or zero-emissions by whoever buys it.

RECs contain specific information about the renewable energy generated, including where, when, at what facility, and with what type of generation. Purchasers of RECs are buying the renewable attributes of those specific units of renewable energy, which helps offset conventional electricity generation in the region where the renewable generator is located. Green-e Energy Certified RECs are not sold more than once or claimed by more than one party, and since they are sold on the voluntary market, they cannot count towards a state's renewable-energy mandate. [6]

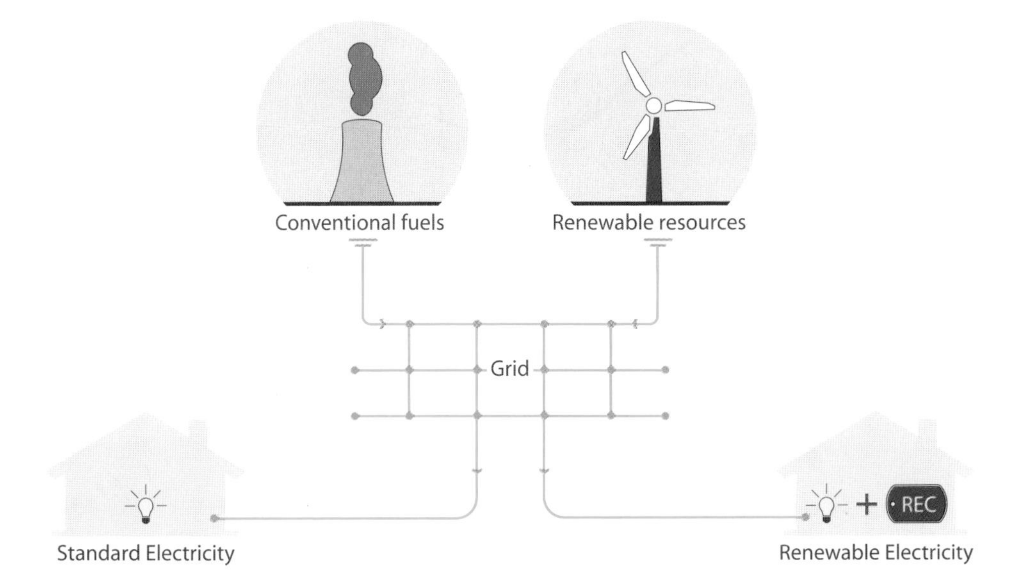

What is Energy Star?

ENERGY STAR® is a joint program of the U.S. Environmental Protection Agency and the U.S. Department of Energy helping us all save money and protect the environment through energy efficient products and practices.
Americans, with the help of ENERGY STAR®, saved enough energy in 2008 alone to avoid greenhouse gas emissions equivalent to those from 29 million cars — all while saving $19 billion on their utility bills.

For Home

Energy efficient choices can save families about a third on their energy bill with similar savings of greenhouse gas emissions, without sacrificing features, style or comfort. ENERGY STAR helps you make the energy efficient choice.

- If looking for new household products, look for ones that have earned the ENERGY STAR. They meet strict energy efficiency guidelines set by the EPA and US Department of Energy.
- If looking for a new home, look for one that has earned the ENERGY STAR.
- If looking to make larger improvements to your home, EPA offers tools and resources to help you plan and undertake projects to reduce your energy bills and improve home comfort.

To earn the ENERGY STAR, a home must meet strict guidelines for energy efficiency set by the U.S. Environmental Protection Agency. These homes are at least 15% more energy efficient than homes built to the 2004 International Residential Code (IRC), and include additional energy-saving features that typically make them 20–30% more efficient than standard homes.

For Business

Because a strategic approach to energy management can produce twice the savings for the bottom line and the environment as typical approaches, EPA's ENERGY STAR partnership offers a proven energy management strategy that helps in measuring current energy performance, setting goals, tracking savings, and rewarding improvements.

The US EPA's ENERGY STAR program has developed energy performance rating systems for several commercial and institutional building types and manufacturing facilities. These ratings, on a scale of 1 to 100, provide a means for benchmarking the energy efficiency of specific buildings and industrial plants against the energy performance of similar facilities. The ratings are used by building and energy managers to evaluate the energy performance of existing buildings and industrial plants. The rating systems are also used by EPA to determine if a building or plant can qualify to earn ENERGY STAR® recognition. [7]

How does it affect us?

Ozone is a gas that is naturally present in our atmosphere. It is made up of 3 oxygen atoms and has a chemical formula O_3. It is created by a sunlight driven reaction between oxygen molecules and it resides in the upper part of atmosphere called stratosphere.

This ozone layer shields us from harmful ultraviolet radiations (UV-B) from the sun. If not absorbed, UV-B would reach the earth surface in amounts that are harmful to life form. The other UV radiation (UV-A), which is not absorbed significantly by ozone, causes premature aging of the skin.

Certain industrial processes and consumer products result in the emission of "halogen source gases" to the atmosphere. These gases bring chlorine and bromine to the stratosphere, which cause depletion of the ozone layer. The source gases that contain only chlorine, fluorine and carbons are called "chlorofluorocarbons", usually abbreviated as CFC's.

Montreal protocol:

In 1987 an international agreement known as the "Montreal protocol on substances that Deplete the Ozone Layer" was signed. This Protocol establishes legally binding controls on the national production and consumption of ozone-depleting gases.

In compliance with the Montreal Protocol the production of CFC in U.S. ended in 1995. Under this Protocol refrigerants with nonzero ODP will be phased out by 2030 in developing countries. This includes CFC's and HCFC's. [8]

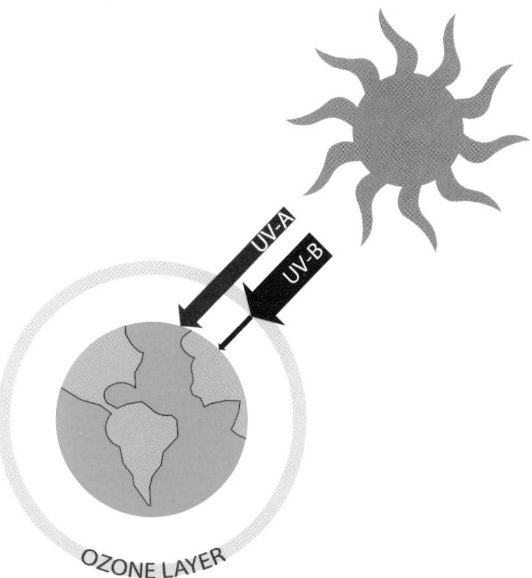

- ***Global Warming:***
  Global warming is the increase in temperature of the Earth's air and oceans since the mid 20th century. The global surface temperature increased by 1°F during the last century and is projected to increase further.
- ***The Clean Air Act:***
  The Clean Air Act is the law that defines EPA's responsibilities for protecting and improving the nation's air quality and the stratospheric ozone layer.

What are refrigerants?

A refrigerant is a fluid which is used in cooling machines like refrigerators and air-conditioners.

Refrigerants fall into following categories:

1) **CFC:** Chlorofluorocarbons cause depletion of stratospheric ozone layer and have higher ozone depletion potential. (ODP)

   HCFC- Hydrochlorofluorocarbons cause significantly less depletion of stratospheric ozone layer than CFC's.

2) **HFC:** Hydrofluorocarbons do not deplete the ozone layer but have a high global warming potential. (GWP)

3) **Non-Halogen group called Natural Refrigerants:** These are environmentally benign to the atmosphere e.g. Carbon dioxide, water, ammonia, hydrocarbon and air.

1 Ozone depleting

O_3

CFC

HCFC

2 Global warming

HFC

3 Environmentally benign

CO_2
H_2O
NH_3
HC
air

- *ODP:*
  Ozone Depletion Potential is a number that refers to the amount of ozone depletion caused by a substance.
- *GWP:*
  Global warming Potential is a number that refers to the amount of global warming caused by a substance.
- *The EPA and WMO (World Meteorological Organization) publishes ODP and GWP values.*

Intent:

To reduce stratospheric ozone depletion and support early compliance with the Montreal Protocol.

Approach:

Refrigerants used in HVAC systems can cause significant damage to earths protective ozone layer if they are released in air.

LEED mandates zero use of CFC based refrigerants in HVAC and refrigeration system.

When reusing existing building HVAC equipments, it requires a comprehensive CFC phase out within 5 years of project completion.

To obtain LEED credits a project needs to do the following:

1) Use no refrigerants
2) Install HVAC systems that use refrigerants with zero or low Ozone Depletion Potential (ODP) and minimal Global Warming Potential(GWP)
   Minimize refrigerant leakage rate
3) Do not use fire suppression systems that contain ozone depleting substances such as CFC's, HCFC's or Halons.

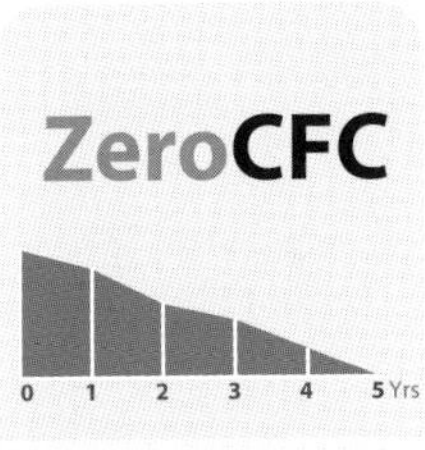

Zero CFC and 5 year phaseout

No refrigerants

Minimum refrigerant leakage

No fire suppression system with halons

- ***Leakage Rate:***
  Leakage rate is the speed at which an appliance loses refrigerant, measured between refrigerant charges or over 12 months, whichever is shorter.

Intent:

To provide ongoing accountability of building energy consumption over time.

Approach:

Measuring and verifying the amount of energy consumed by the building after occupancy, optimizes the performance of the building, and reduces the economic and environmental impact associated with its energy-related systems.

The IPMVP Volume III presents four options for measurement and verification (M&V) plan. Out of these Option B and Option D are appropriate for LEED M&V.

Option B: Energy Conservation Measure Isolation

This method determines savings by measuring energy use and operating parameters of system to which measure was applied, separate from rest of facility.

This method is suitable for small and/or simple building.

Option D: Whole-Building Simulation

This method determines savings at whole-building or system level by measuring energy use and comparing it with baseline.

This method is suitable for building with large number of energy conserving measures.

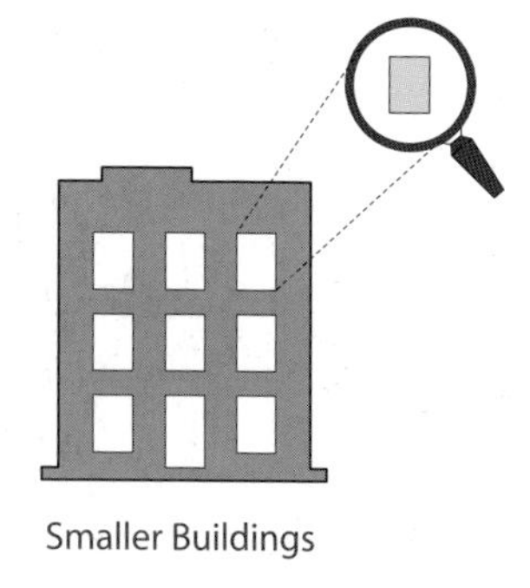
Smaller Buildings

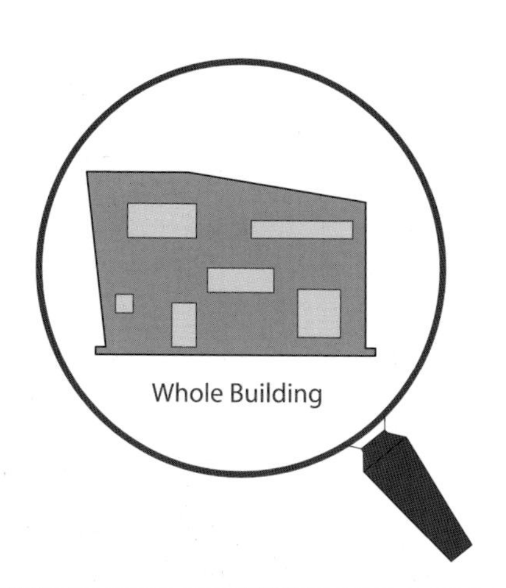
Whole Building

○ *IPMVP:*

International Performance Measurement and Verification Protocol. IPMVP Inc is a nonprofit organization whose vision is a global marketplace that properly values energy and water efficiency. [9]

Building construction and operation generate huge amount of waste. This generated waste ends up in landfills and incinerators, and creates environmental impacts. Thus it is important that the quantity of waste generated is reduced and recycled.

LEED addresses this environmental concern by promoting following measures:

- Practicing waste reduction
- Reducing waste at its source
- Reusing and recycling
- Selecting sustainable materials

This chapter explains eight strategies in detail to achieve all the above mentioned intents.

Intent:
To facilitate the reduction of waste generated by building occupants.

Approach:
By providing recycling facilities in the building, a significant amount of waste can be diverted from going into the landfills. Recycling saves money as the tipping cost is significantly reduced.

LEED requires provision of an accessible area for collection and storage of non-hazardous materials for recycling. Materials must include at a minimum - corrugated cardboard, paper, glass, metals and plastic.

The recycling can be commingled (all in one container) or in different containers. The advantage of commingled recycling is that people and contractors are more likely to use it, and it also requires less space.

In large projects, employing cardboard balers, aluminum can crushers, recycling chutes, and other waste management technology can improve recycling program and reduce volume of recycled material and required space.

Collection of waste materials

Recycle: Cardboard, paper, glass, metal & plastic

○ ***Tipping fees:***
The fees charged by a landfill for disposal of waste are called tipping fees. They are typically quoted per ton.

Intent:
To extend the life cycle of the building stock, preserve natural and cultural resources and reduce waste.

Approach:
The most effective strategy to minimize environmental impact is reusing an existing building rather than building a new one.

Some benefits of reusing an existing building are:
- Reduction in the cost of construction.
- Reduction in the energy used for demolition process.
- Reduction in environmental impacts associated with extraction, manufacturing and transportation.

LEED recommends maintaining a substantial percentage of the following existing building components:
1) **Building structure** - structural floor, roof decking, exterior skin and framing excluding windows and non-structural roofing materials.
2) **Interior-non structural elements** - interior walls, doors, floor coverings and ceiling systems.

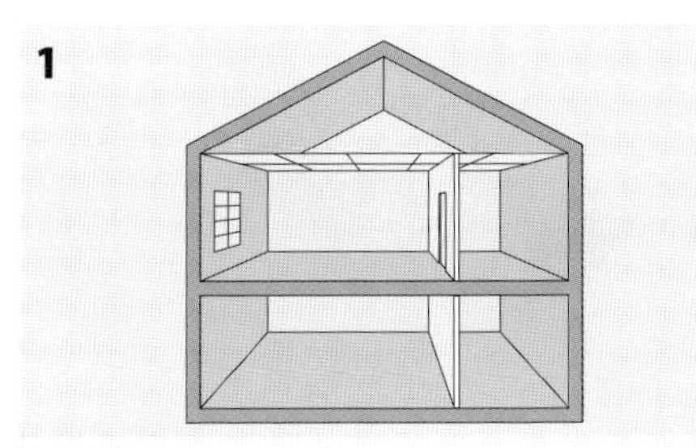

Building structure

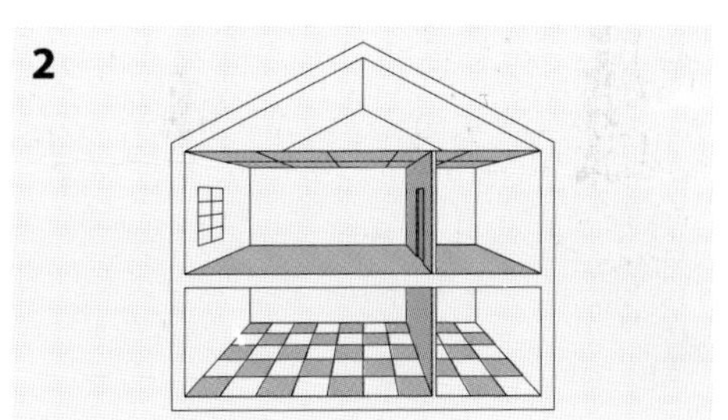

Interior non-structural elements

- *The percentage of materials reused is calculated on sq. ft. basis.*
- *Hazardous materials are excluded from this calculation.*

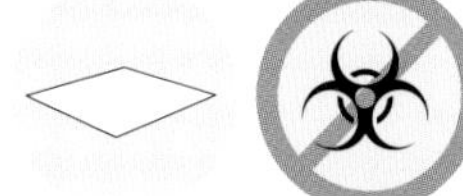

○ ***Reused area:***
Reused area is the total area of the building structure, core, and envelope that existed in the prior condition and remains in the complete design.

Intent:

To divert construction and demolition debris, from disposal in landfills and incinerators.

Approach:

Construction waste management is very crucial part of sustainable building. Waste management means minimizing the construction waste or demolition debris (C&D), which leaves the site. As per EPA this C&D debris contribute to 40% of landfill waste each year. [1]

LEED credit can be obtained by doing the following:

1) Recycling, salvaging or donating a percentage of nonhazardous construction and demolition debris.
2) Developing and implementing a construction waste management plan that does the following:
   - Identifies materials to be diverted from disposal
   - Determines if the materials will be sorted on site or commingled.

The following materials do not contribute to this LEED credit:

- Excavated soil – e.g. fill dirt
- Land clearing debris e.g. – tree stumps, vegetation
- Hazardous materials e.g. – sheet rock, PCB, ACM

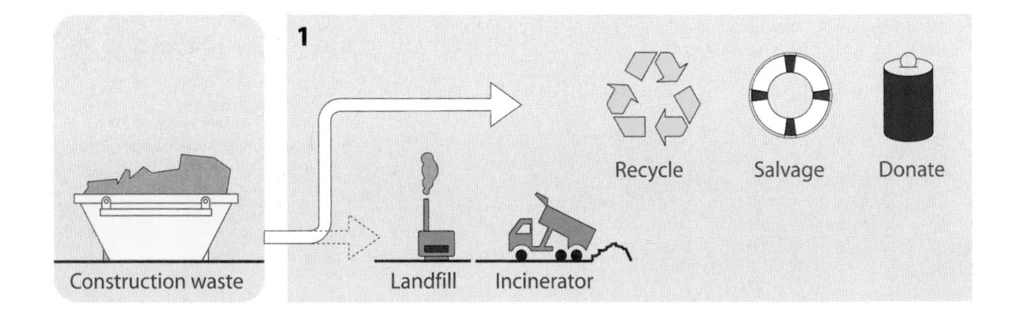

- *Calculations of waste diverted from disposal are done based on weight OR volume.*

- ***PCB:***
  Polychlorinated biphenyl (PCB) is a hazardous organic compound which was used in commercial and industrial applications because of its slow degradation and chemical stability.
- ***ACM:***
  Asbestos-containing materials (ACM). Asbestos is a hazardous fibrous mineral used for its fire resistance, noise insulation and electrical insulation properties.

Intent:
To reuse building materials & products to reduce demand for virgin materials and reduce waste.

Approach:
Many building materials can be salvaged, refurbished and reused. Reusing materials reduce the need for landfill space and prevents environmental impacts of water and air pollution.

LEED recommends using a percentage of salvaged, refurbished or reused materials in the project.

The strategies to reduce waste as per EPA are:
1) Reduce - the amount and toxicity of trash you throw away
2) Reuse - containers and products
3) Recycle - as much as possible
4) Buy recycled - products with recycled content [2]

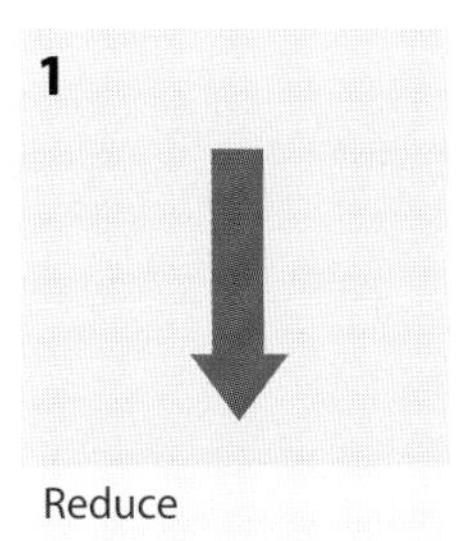

Reduce

Reuse

Recycle

Buy recycled

- *Mechanical, Electrical and Plumbing components, and special items like elevators cannot be included for the calculation.*
- *Calculations of the material reused are done based on the cost of materials.*

- ***Source Reduction:***
  Source reduction, is the practice of designing, manufacturing, purchasing, or using materials (such as products and packaging) in ways that reduce the amount or toxicity of trash created. Reusing items is another way to stop waste at the source because it delays or avoids that item's entry in the waste collection and disposal system. [3]
- ***Salvaged materials:***
  Salvaged materials are construction materials recovered from existing building or site. Some of the typically salvaged materials include structural beams and posts, flooring, doors, cabinetry, bricks and decorative items.

Intent:

To increase the demand of products that incorporate recycled content materials.

Approach:

Products which comprise of recycled content, reduce the need to extract and process virgin materials. They also reduce the volume of solid waste which otherwise would have been generated. LEED encourages the use of such products which have recycled contents in them.

There are two types of recycled contents:

1. **Pre-consumer:** These are materials diverted from waste stream during manufacturing. Examples includes; sawdust, bagasse, walnut shells, culls, trimmed materials, overissue publications and obsolete inventories.
2. **Post–consumer:** These are the materials that can be no longer used for their main purpose and are discarded by the consumers. Examples include; construction and demolition debris, materials collected through recycling programs, discarded products (e.g. furniture, cabinetry, decking), and landscaping waste (e.g. leaves, grass clippings, tree trimmings).

For steel products where no recycled content information is available, LEED allows project teams to assume its recycled content to be 25% post-consumer.

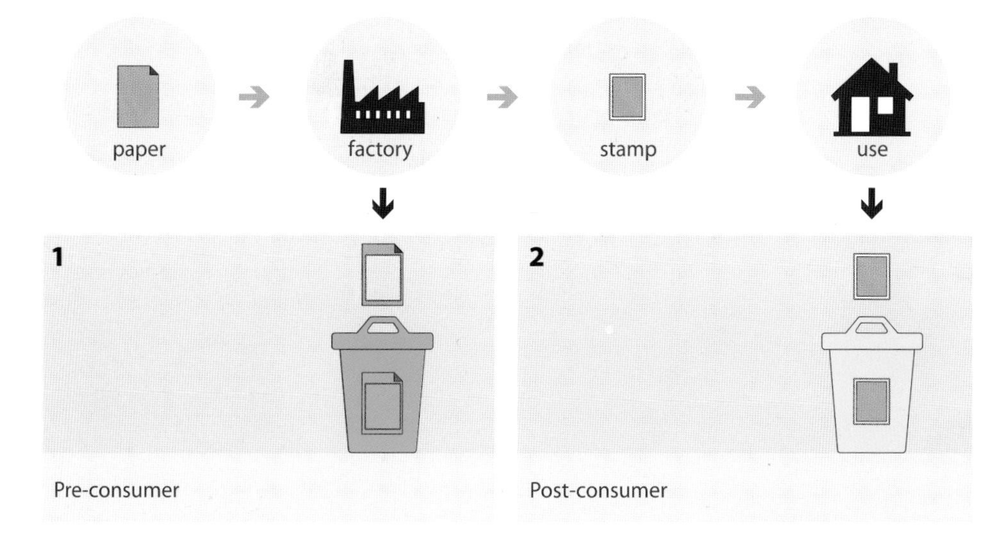

○ ***Fly Ash:***
Fly ash is the solid residue obtained from incineration process. This can be used as a substitute for Portland cement in concrete.

Intent:

To increase the demand for building products which are extracted and manufactured within the region, and thus reducing the environmental impact caused due to transportation of those materials.

Approach:

Using regional building materials has following benefits:

- Reduces transportation activity and there by the pollution caused by it.
- Reduces the cost of transportation.
- Reduces the energy required for transportation.

To obtain this LEED credit it is required that a percentage of the materials used on site, are extracted, harvested or recovered as well as manufactured within 500 miles radius of the site.

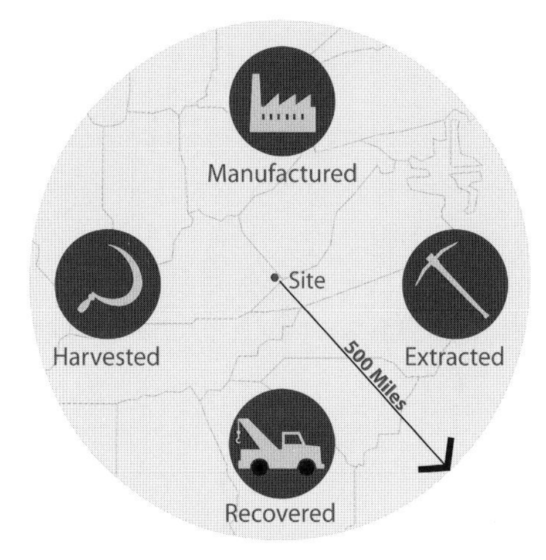

○ ***Manufacturing:***
Manufacturing is the final assembly of various materials into a building product, which is then furnished and installed by the trade workers.

Intent:

To reduce the use and depletion of finite raw materials by replacing them with rapidly renewable materials.

Approach:

The most conventional building materials require large quantity of natural resources, time and money to produce them. Where as rapidly renewable materials are eco-friendly and have less environmental impact.

To obtain a LEED credit it is required that a percentage of building materials and products used on site, are rapidly renewable.

Following are the few examples of rapidly renewable materials:

1

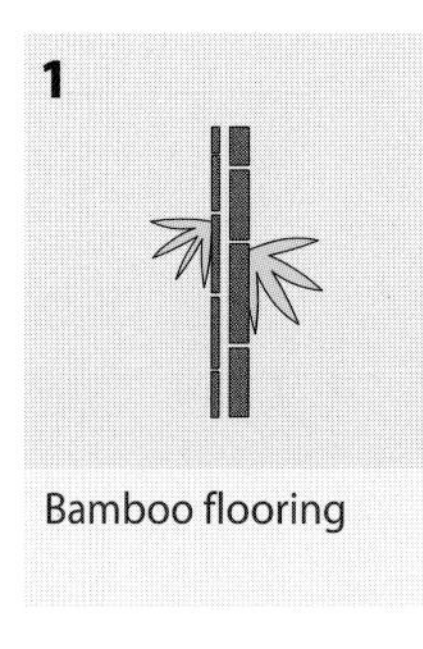

Bamboo flooring

2

Wool carpeting

3

Cotton insulation

4

Wheat board

5

Cork flooring

6

Linoleum flooring

○ ***Rapidly renewable materials:***
Rapidly renewable materials are agricultural products, both fiber and animal, that take 10 years or less to grow or raise and can be harvested in a sustainable fashion.

Intent:

To encourage an environmentally responsible forest management.

Approach:

An irresponsible forest management is harmful to the forest, its wildlife, soil, water and its air. The Forest Stewardship council (FSC) standards ensure sustainability and integrity of the forest ecosystem. [4]

Projects achieve a LEED credit if they use a percentage of wood-based materials which are certified by the FSC. This certified wood comes with a Chain of custody (CoC) certificate.

- ***Chain of Custody (CoC):***
  Chain of Custody (CoC) is a procedure which keeps track of the journey of a product from its starting point of harvesting or extraction, to its end point of distribution. [5]

On an average, Americans spend 90% of the time indoors, hence quality of the indoor environment has a significant influence on the health, productivity and quality of life. Environmental Protection Agency (EPA) reports that the pollutant levels of the indoor environment may be 2 to 5 times, and occasionally even 100 times higher than the outdoor levels. [1]

This chapter addresses environmental concerns relating to the indoor environmental quality; health, safety and comfort of the occupants; energy use; effective air change and air contaminant management. Following are few strategies that address these concerns:

- Improving Ventilation
- Managing Air Contaminants
- Specifying Less Harmful Materials
- Allowing Occupants to Control Desired Settings
- Providing Daylight and Views.

This chapter elaborates above mentioned strategies which enhance the indoor environment and optimize interior spaces for building occupants.

Intent:

To establish a minimum indoor air quality (IAQ) performance to enhance the air quality inside the building.

Approach:

A good Indoor Air Quality (IAQ) in building improves productivity, comfort and well being of its occupants.

There are three basic strategies for ventilating buildings:

1) Mechanical ventilation (i.e. active ventilation)
2) Natural ventilation(i.e. passive ventilation)
3) Mixed-ventilation (i.e. both mechanical and natural ventilation)

LEED requires that the ventilation system meet the requirements mentioned in ASHRAE Standards.

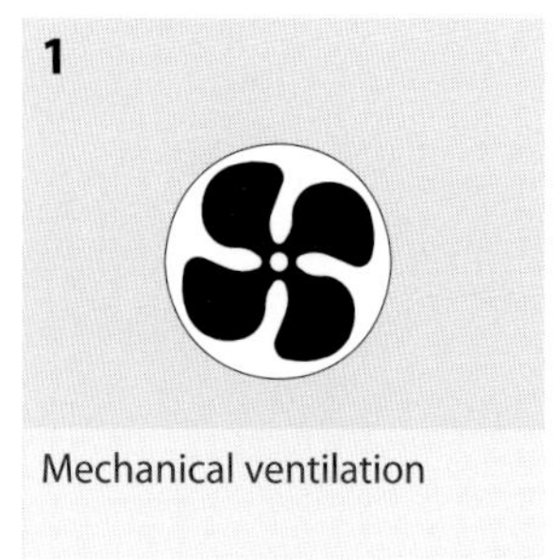

Mechanical ventilation

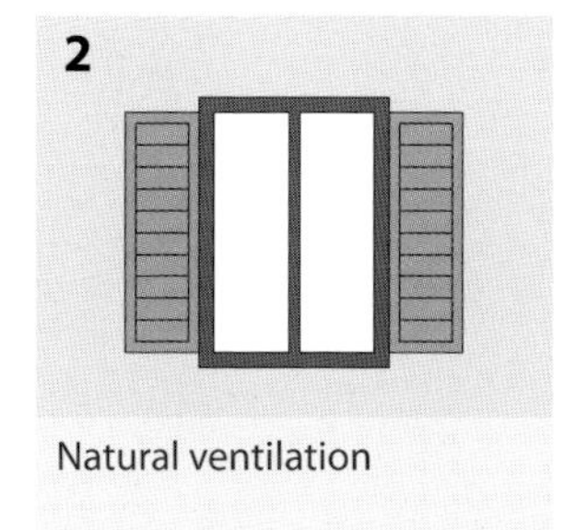

Natural ventilation

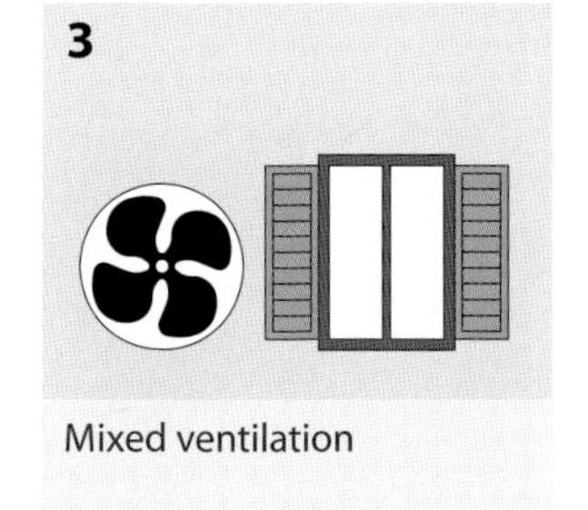

Mixed ventilation

Ventilation:

Ventilation is the process of supplying air to, or removing air from a space for the purpose of controlling air contaminant levels, humidity, or temperature within the space.

Intent:

To minimize exposure of building occupants, indoor surfaces, and ventilation systems to Environmental Tobacco Smoke (ETS).

Approach:

Environmental tobacco smoke, or secondhand smoke, comprises of the smoke given off by lighted tobacco products and smoke exhaled by smokers. Tobacco smoke contains various carcinogenic chemicals. Exposure to this smoke may cause various lung and heart diseases. Thus it is of prime importance that smoking be prohibited in the building and should be limited to designated outdoor areas.

○ ***Environmental tobacco smoke (ETS):***

Environmental tobacco smoke or second hand smoke, consist of airborne particles emitted from the burning end of cigarettes, pipe, and cigars, and is exhaled by smokers. These particles contain about 4,000 different compounds, up to 50 of which are known to cause cancer. [2]

Intent:

To make ventilation system, capable of monitoring outdoor air.

Approach:

In order to achieve a better indoor air quality (IAQ), the carbon dioxide (CO_2) concentrations should be measured. The CO_2 levels are an indicator of air-change effectiveness. An increase in CO_2 level suggests inadequate ventilation and a possible buildup of indoor air pollutants.

LEED recommends installing permanent monitoring systems to ensure the minimum ventilation requirement as per the design. The CO_2 sensors are placed in the breathing zone.

The breathing zone is the region within an occupied space between 3 feet and 6 feet above the floor and more than 2 feet from the walls or fixed air conditioning equipment.

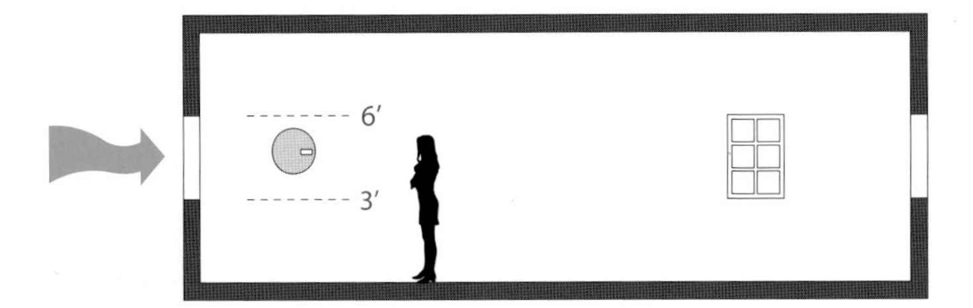

- ***Outdoor air:***
  Outdoor air is the ambient air that enters a building through a ventilation system, either through intentional openings for natural ventilation or by infiltration. (ASHRAE 62.1-2007)

Construction IAQ Management – During Construction & Before Occupancy

Intent:

To reduce indoor air quality (IAQ) problems resulting from construction or renovation and promote the comfort and well-being of construction workers and building occupants.

Approach:

During construction or demolition, contaminants are inevitably introduced in building interiors. If this issue is not addressed during construction and before occupancy, the contaminants can cause a very poor IAQ extending over the lifetime of the building. Thankfully there are IAQ management strategies which, if implemented during construction and before occupancy, can minimize the problem.

LEED recommends following strategies for construction IAQ management:

1) Meet or exceed control measures of SMACNA IAQ guidelines for occupied buildings under construction.
2) Protect, stored on-site or installed absorptive materials from moisture damage.
3) During construction, use filtration media with minimum efficiency reporting value (MERV) of 8 at each return air grille.

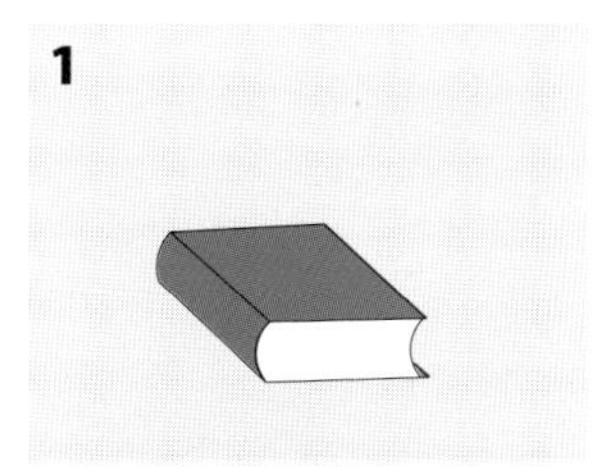

Meet or exceed control measures of SMACNA

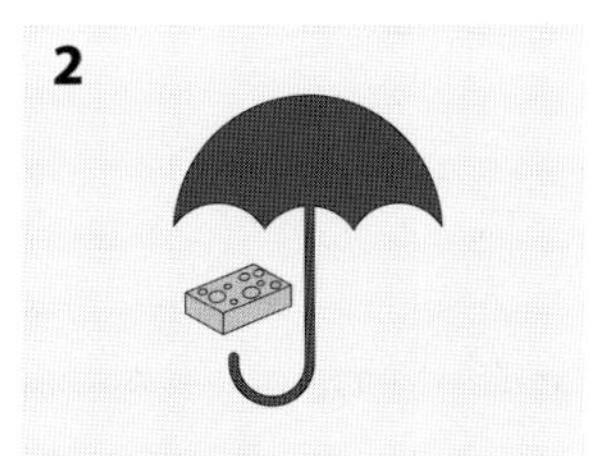

Protect absorptive materials

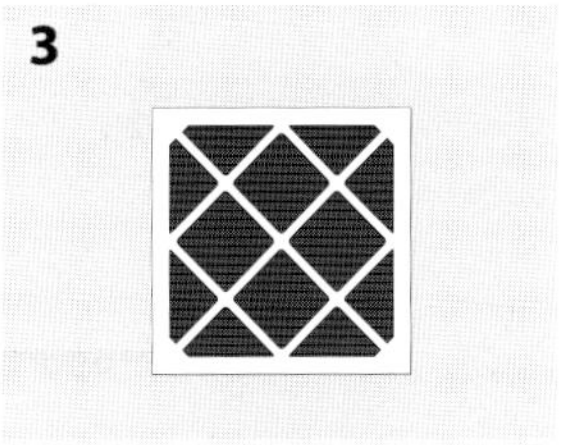

During construction use MERV 8 filters

- ***SMACNA:***
  SMACNA (Sheet Metal and Air Conditioning Contractors' National Association, Inc.) - This is an international organization that developed guidelines for maintaining healthful indoor air quality during demolitions, renovations and construction. [3]
- ***Bake-out:***
  A process used to remove VOC's from a building by elevating the temperature in a fully furnished and ventilated building before human occupancy.

Intent:

To reduce the quantity of odorous, irritating and harmful indoor air contaminants.

Approach:

Preventing indoor environmental quality problems are effective and less expensive than solving them later. The best way to tackle these problems is to specify materials that release less harmful chemical compounds.

Many building products contain volatile organic compounds (VOC's). These VOC's react with sunlight and nitrogen oxides in the atmosphere to form ground level ozone. This chemical contributes to smog formation, air pollution, and is harmful to human health, crops and ecosystem.

Adhesives, paints, carpets, composite wood products and furniture with low level of potentially irritating off-gassing can reduce its exposure to occupants and harm caused by it.

LEED recommends that various building products comply with their respective material standards, so that they do not exceed permissible VOC limits.

Adhesives & Sealants

Adhesives, Sealants, and Sealant Primers used on the interiors: **South Coast Air Quality Management District Rule #1168.**

Aerosol Adhesives: **GS-36**

Flooring Systems

All carpets used in interiors: **Carpet and Rug Institute's Green Label Plus Program.**

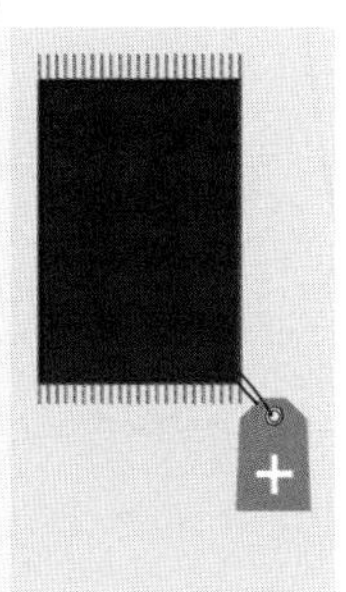

Paints & Coatings

Non-Flat Paints and coatings used on the interiors: **Green Seal Standard GS-11**

Anti-Corrosive, anti Rust Paints: **GC-03**

Architectural Coatings: Clear wood finishes, floor coatings, stains, sealers, shellacs: **SCAQMD Rule 1113**

Composite Wood & Agrifiber Products

Composite wood and Agrifiber products (particleboard, MDF, plywood, wheat board, strawboard, panel substrates, door cores) used in interiors: **NO** added **U**rea-**Fo**rmaldehyde resins.

NO UFo

- ***Off-gassing:***
  Off-gassing is the emission of volatile organic compounds (VOCs) from synthetic and natural products.

Intent:

To minimize the exposure of building users to hazardous particulates and chemical pollutants.

Approach:

It is essential to reduce or eliminate human contact with airborne chemicals and particles.

LEED suggests following strategies to control and minimize the entry of pollutants into buildings:

1) **Entryway systems:** Incorporating permanent entryway system, at exterior entrances. The entryway system like grilles, mats etc capture and remove particles from shoes and keep the interiors clean.
2) **Exhaust hazardous gases:** Exhaust spaces which have hazardous gases or chemicals present in them (e.g. garages, laundry areas, labs, print rooms etc.). For effective removal of contaminants, deck to deck partitions and self closing doors must be incorporated.
3) **Filtration Media:** Install new air filtration media before occupancy, which should have (Minimum efficiency reporting Value) MERV 13 or heigher
4) **Containment of waste:** Provide closed container for storage for disposal as per the regulations.

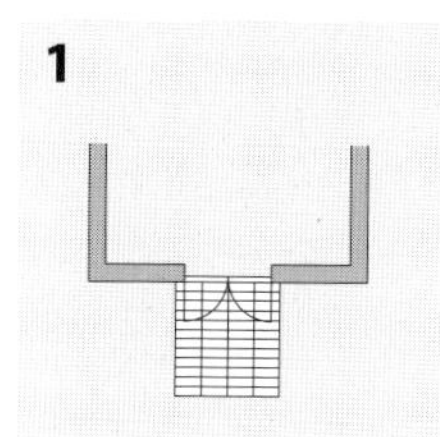

Entryway system

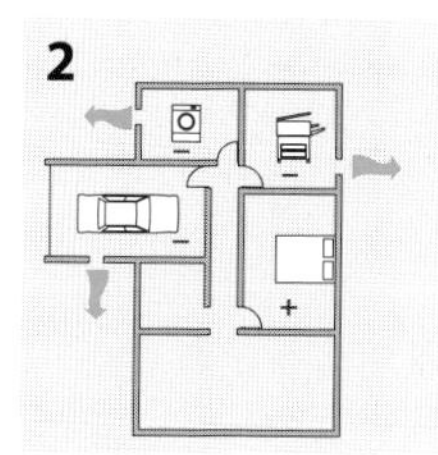

Exhaust hazardous gases

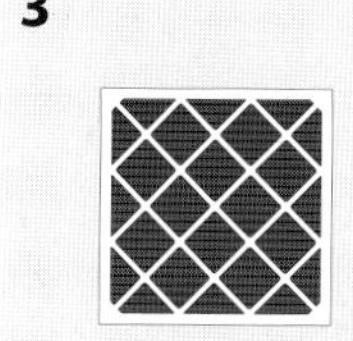

Filteration media

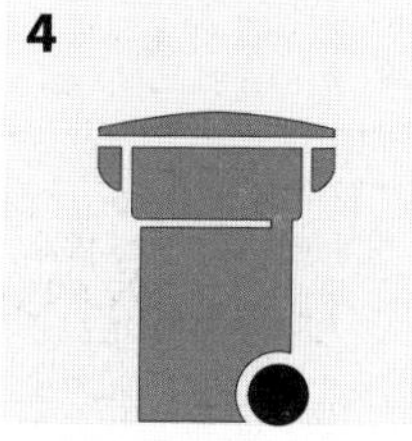

Containment of waste

○ ***Particulates:***
Solid particles or liquid droplets in the atmosphere.

○ ***MERV:***
Minimum efficiency reporting value (MERV) is a filter rating system established by ASHRAE. MERV categories range from 1(very low efficiency) to 16 (very high efficiency).

Intent:

To provide a high level of lighting system control and thermal comfort systems control, by building occupants.

Approach:

Effective lighting is essential for human comfort, productivity and communication.

Providing individual controls for lighting and thermal comfort increases the comfort level of occupants by enabling them to customize their workspace as per their individual needs.

LEED recommends individual lighting controls for individual as well as shared multi occupant spaces.

Individual lighting controls

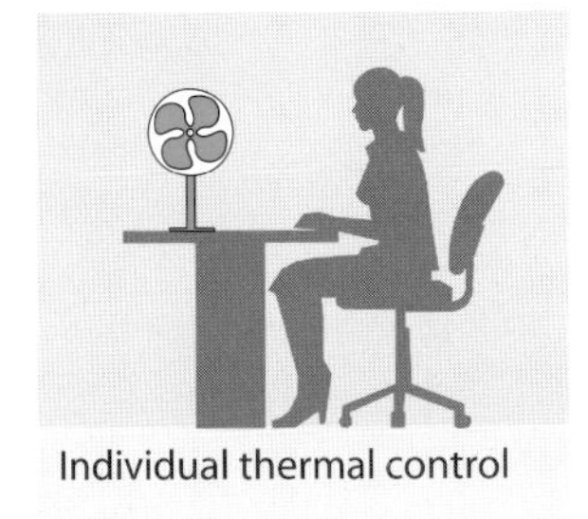

Individual thermal control

Lighting and thermal comfort for shared spaces

○ ***Controls:***

Controls are operating mechanisms that enable a person to turn on or off devices (e.g. lights, heaters) or adjust systems within a range (e.g. lighting, temperature).

Intent:

To provide a comfortable thermal environment for the building occupants, and to provide for its assessment over time.

Approach:

It is necessary to maintain an acceptable level of thermal comfort for building occupants. People who are comfortable are more productive and happier. A comfortable environment boosts productivity and speed, which means efficient use of equipment like computers, and hence energy saving.

LEED point can be obtained, if the building HVAC (Heating Ventilating and Air Conditioning) system design, and the building envelope meet the requirements of ASHRAE standard 55-2004, Thermal Environmental Conditions for Human Occupancy.

An additional LEED point can be obtained, if a thermal comfort survey is conducted of the building occupants within 6 to18 months of occupancy. If 20% or more occupants are dissatisfied with the thermal comfort, a plan for corrective action should be developed.

Design HVAC & building envelope as per standards.

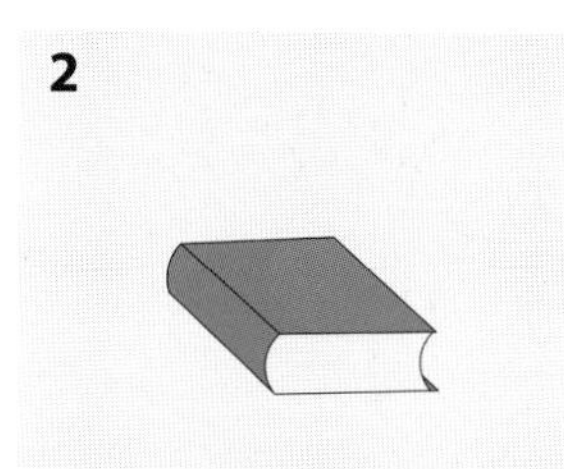

Comply with ASHRAE Standard 55-2004

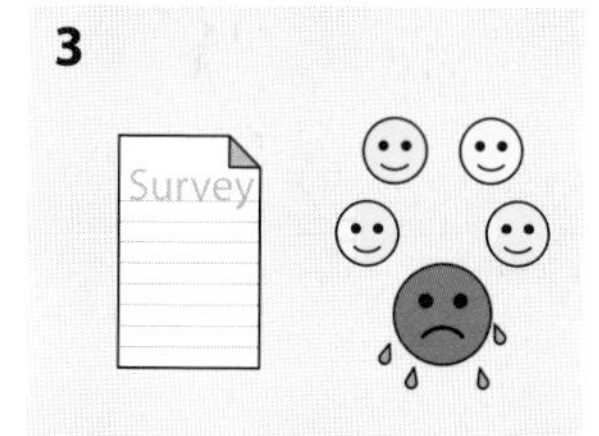

Thermal comfort survey after 6 to 18 months of occupancy

○ ***Predicted mean vote:***
It is an empirical equation for predicting the mean vote on a rating scale of thermal comfort of a large population of people exposed to a certain environment. (Used in verification of thermal comfort design).

Intent:

To connect building users with outdoors by means of daylight and views.

Approach:

Daylighting decreases the need for artificial lighting and hence reduces the energy consumption. Daylight can be introduced in the building by following ways:

- Shading devices
- Light shelves
- Courtyards
- Atriums
- Window glazing

It is important to consider the following while designing:

- Orientation of building
- Window size and spacing
- Glass selection
- Reflectance of interior finishes
- Location of interior walls

One of the very common failures in daylighting strategies is glare control. The glare can be controlled by following strategies:

1) Fixed exterior shading devices like fins
2) Exterior light shelves
3) Interior light shelves
4) Interior louvers and blinds
5) Fritted glass
6) Electronic blackout glazing.

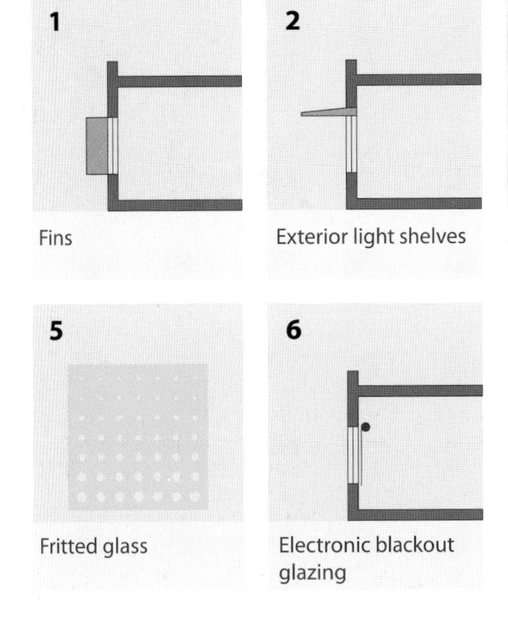

1 Fins

2 Exterior light shelves

5 Fritted glass

6 Electronic blackout glazing

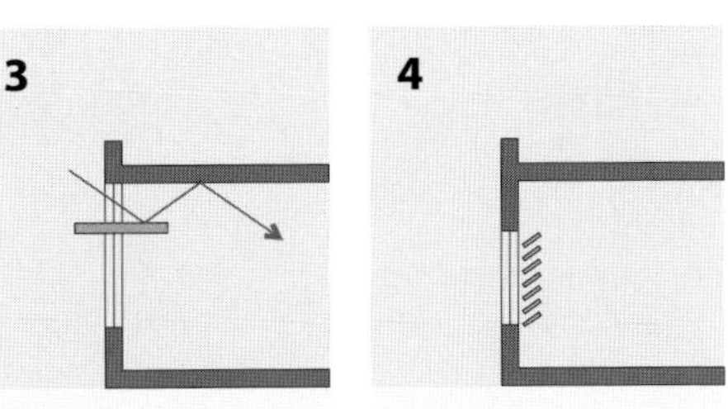

3 Interior light shelves

4 Louvers & blinds

○ ***Glare:***
Glare is any excessively bright source of light within the visual field that creates discomfort or loss of visibility.

Sustainable design strategies and measures are constantly evolving and improving. New technologies are introduced continuously and modern scientific research influence buildings. This category recognizes projects for innovative building features and sustainable building practices.

Intent:

To provide design team and projects an opportunity to achieve exemplary performance above the requirements set by the LEED rating system, and/ or innovative performance in categories not specifically addressed by LEED rating system.

Approach:

This LEED credit can be achieved by following two ways:

1) **Exemplary Performance**
   These credits are awarded for doubling the credit requirement and/or achieving the next incremental percentage threshold.

2) **Innovation in Design**
   These credits are awarded for comprehensive strategies which demonstrate quantifiable environment benefits.

Following are some examples of innovative performance credits awarded to LEED certified projects:

- Education outreach program
- Green housekeeping
- Use of high volume fly ash
- Low-emitting furniture and furnishings
- Organic landscaping/ integrated pest management program

Innovation in design ideally begins at a projects conception, but it can enter at any step of the process and come from any member of the project team.

Intent:

To support and encourage the design integration required by LEED to streamline the application and certification process.

Approach:

LEED AP's have the expertise to design a building as per LEED standards and to coordinate the documentation process necessary for the LEED certification.

Appropriate candidates for accreditation are as follows:

- Architects
- Engineers
- Consultants
- Owners
- Anyone who has a strong interest in sustainable building design.

LEED AP's should champion the project's LEED application and be an integral member of the project team. LEED AP's should also help with synergies among prerequisites and credits.

Projects can earn one credit if at least one principal participant of the project team is a LEED Accredited Professional.

○ *LEED AP:*

LEED Accredited Professionals (APs) are individuals who have successfully completed the LEED professional accreditation exam.

Since some environmental issues are unique in a particular region USGBC regional councils have identified distinct environmental zones within their region and allocated 6 additional credits to encourage design teams to focus on regional priorities. A project that earns a Regional Priority credit automatically earns one point in addition to any points awarded for that credit.

REGIONAL PRIORITY

Intent:

To encourage design teams to achieve credits that address geographically specific environmental priorities.

Approach:

Regional priority credits are listed by state and zip code. The credits are specific to a projects zip code and the list is available on the USGBC website.

One point is awarded for each Regional Priority credit achieved; maximum four RP credits can be achieved. These credits are not available to projects outside of the U.S.

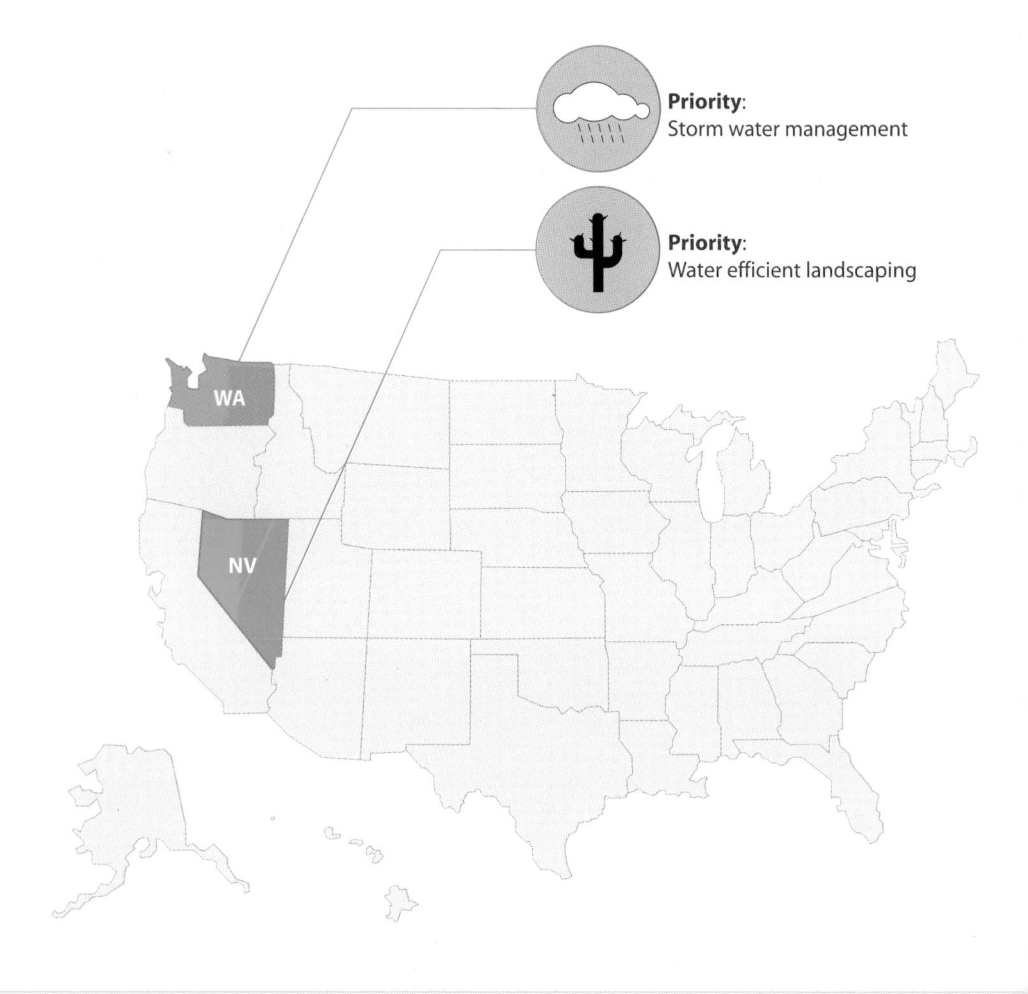

○ *Note:*

Regional variation is different from Regional Priority.

EXAM
• REFERENCE •
INDEX

1 Awareness:

To be eligible to take the LEED Green Associate exam you have to meet any ONE, of the following criteria:

- **Experience** – You must have experience in the form of involvement on a LEED-registered project.
- **Employment** – Current employment (or previous employment) in a sustainable field of work.
- **Engagement** – Engagement in (or completion of) an education program that addresses green building principles.

Your experience must be documented in the form of a letter of attestation from a supervisor, client, project manager, or teacher and must describe your involvement on the job or in the classroom.

2 Understanding:

The LEED AP exam is divided into two parts. The first part is the LEED Green Associate exam, which demonstrates general knowledge of green building practices. The second part is a specialty exam based on one of the LEED Rating Systems.

3 Implementation:

After passing the LEED Green Associate Exam, one can pursue to become a LEED AP with specialization in one or more of the following:

- **Building Design + Construction**
- **Homes**
- **Interior Design + Construction**
- **Neighborhood Development** (available 2010)
- **Operations + Maintenance.**

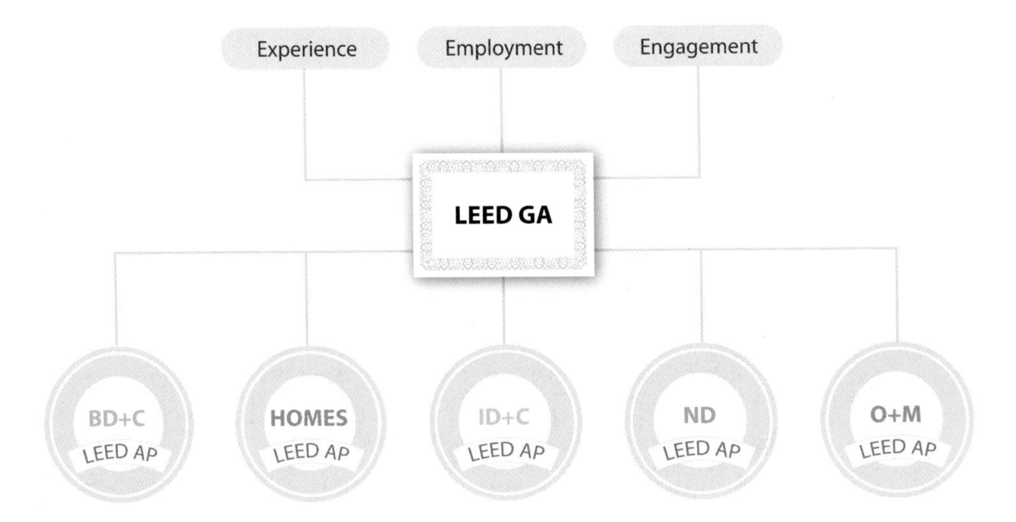

References

Green Building

1 Green Building research. www.usgbc.org (accessed November 2009)
2 Green building basic information. http://www.epa.gov/greenbuilding/pubs/about.htm (accessed October 2009
3 Green building research. www.usgbc.org (accessed November 2009)
4 USGBC; Green Building & LEED Core Concepts Guide, first edition (e-book)
5 National Institute of Standards and Technology (NIST) Handbook 135, 1995 edition.
6 Whole building design guide , www.wbdg.org (accessed November 2009)
7 Don Prowler. Whole building Design. www.wbdg.org/wbdg_approach.php (accessed November 2009)

Leadership in Energy & Environment Design

1 Foundations of LEED.www.usgbc.org (accessed November 2009)
2 About USGBC. www.usgbc.org (accessed November 2009)
3 Intro-what LEED is. www.usgbc.org (accessed November 2009)
4 Process overview. www.gbci.org (accessed November 2009)
5 LEED 2009 Minimum program requirements. www.gbci.org (accessed November 2009)
6 Credit Interpretation Ruling . www.gbci.org (accessed October 2009)
7 USGBC; LEED Green Associate Study Guide (U.S. Green Building Council, 2009)

Sustainable Site

1 Brownfield and land revitalization. http://www.epa.gov/brownfields/index.html (accessed November 2009)
2 Unites States Environmental protection Agency http://www.epa.gov/epahome/aboutepa.htm (accessed November 2009)
3 USGBC, LEED for Homes Rating System (U.S. Green Building Council, 2008)

Water Efficiency

1 USGBC." LEED Certified Project List" www.usgbc.org/LEED/project/certifiedprojectlist.aspx (accessed May 2008)
2 U.S. EPA, Office of Water, Water-Efficient landscaping.2002. www.epa.gov/owm/water-efficiency/final_final.pdf (accessed January 2005)
3 Energy policy act. http://www.gc.energy.gov (accessed October 2009)
4 Watersense. www.epa.gov/WaterSense/ (accessed November 2009)

Energy & Atmosphere

1 US Department of Energy, Office of Energy Efficiency and Renewable energy. "Table 1.1.1 U.S. Residential and Commercial Buildings Total Primary Energy Consumption,2006." 2008 Buildings Energy Data Book. 2008. http://buildingsdatabook.eren.doe.gov (accessed November 2008)
2 About us. http://www.ashrae.org/aboutus/ (accessed November 2009)
3 About ICC. http://www.iccsafe.org/AboutICC/Pages/default.aspx (accessed November 2009)
4 ADA standards for accessible design. http://www.ada.gov/ (accessed November 2009)
5 ISO 14000 essentials http://www.iso.org/iso/iso_14000_essentials (accessed November 2009)
6 Green-e Energy. http://www.green-e.org/getcert_re.shtml (accessed Nov 2009)
7 About Energy star. www.energystar.gov (accessed November 2009)
8 D.W. Fahey, Twenty Questions and Answers about the ozone layer: 2006 Update. http://www.esrl.noaa.gov/csd/assessments/2006/chapters/twentyquestions.pdf (accessed October 2009)
9 History. www.evo-world.org (accessed November 2009)

Material & Resources

1 U.S. Environmental Protection Agency, Office of Solid Waste. Municipal Solid waste Generation, Recycling, and disposal in the United states: Facts and Figures for 2005.2006. http://www.epa.gov/osw/rcc/resources/msw-2005.pdf (accessed November 2008)
2 U.S. Environmental Protection Agency, Reduce, reuse, Recycle, Buy recycled. http://www.epa.gov/region09/waste/solid/reduce.html
3 U.S. Environmental Protection Agency, Reduce & Reuse. http://www.epa.gov/osw/conserve/rrr/reduce.htm (accessed November 2009)
4 Forest Stewardship council, Principles and criteria. http://www.fscus.org/standards_criteria/ (accessed November 2009)
5 Forest Stewardship council , Chain of custody certification http://www.fsc.org/134.html (accessed November 2009)

Indoor Environmental Quality

1 U.S. Environmental Protection Agency. Healthy Buildings, Healthy people: A vision for the 21st century. 2001 http://www.epa.gov/iaq/hbhp/hbhptoc.html (accessed May 2008)
2 Department of Health and Human Services, National Institute of health, National Cancer Institute. Health Effects of exposure to environmental Tobacco smoke- Smoking and Tobacco Control Monograph 10.1999 http://cancercontrol.cancer.gov/tcrb/monographs/10/m10_complete.pdf (accessed may 2008)
3 Sheet metal and air conditioning Contractors' National Association, about SMACNA. http://www.smacna.org/about/ (accessed October 2009)

Index

Bibliography

• LEED Reference Guide for Green Building Design and Construction(U.S. Green Building Council, 2009)
• LEED for Operations & Maintenance Reference Guide-Introduction (U.S. Green Building Council, 2008)
• LEED for Operations & Maintenance Reference Guide-Glossary (U.S. Green Building Council, 2008)
• LEED for Homes Rating System (U.S. Green Building Council, 2008)
• Cost of Green Revisited, by Davis Langdon (2007)
• Sustainable Building Technical Manual: Part II, by Anthony Bernheim and William Reed (1996)
• The Treatment by LEED® of Environmental Impact of HVAC Refrigerants (LEED Technical and Scientific Advisory Committee, 2004)
• Guidance on Innovation & Design (ID) Credits (US Green Building Council, 2004)
• Guidelines for CIR Customers (US Green Building Council, 2007)
• Energy Performance of LEED® for New Construction Buildings: Final Report, by Cathy Turner and Mark Frankel (2008)
• Foundations of the Leadership in Energy and Environmental Design Environmental Rating System: A Tool for Market Transformation (LEED Steering Committee, 2006)
• AIA Integrated Project Delivery: A Guide (www.aia.org)
• Review of ANSI/ASHRAE Standard 62.1-2004: Ventilation for Acceptable Indoor Air Quality, by Brian Kareis (www.workplacegroup.net)
• Best Practices of ISO - 14021: Self-Declared Environmental Claims, by Kun-Mo Lee and Haruo Uehara (2003).
• Bureau of Labor Statistics (www.bls.gov)
• International Code Council (www.iccsafe.org)
• Americans with Disabilities Act (ADA): Standards for Accessible Design (www.ada.gov)
• GSA 2003 Facilities Standards (General Services Administration, 2003)
• Guide to Purchasing Green Power (Environmental Protection Agency, 2004)
• Green Building & LEED Core Concepts Guide•, 1st Edition (US Green Building Council, 2009)